Lecture Notes in Statistics

Vol. 1: R. A. Fisher: An Appreciation. Edited by S. E. Fienberg and D. V. Hinkley. XI, 208 pages, 1980.

Vol. 2: Mathematical Statistics and Probability Theory. Proceedings 1978. Edited by W. Klonecki, A. Kozek, and J. Rosiński. XXIV, 373 pages, 1980.

Vol. 3: B. D. Spencer, Benefit-Cost Analysis of Data Used to Allocate Funds. VIII, 296 pages, 1980.

Vol. 4: E. A. van Doorn, Stochastic Monotonicity and Queueing Applications of Birth-Death Processes. VI, 118 pages, 1981.

Vol. 5: T. Rolski, Stationary Random Processes Associated with Point Processes. VI, 139 pages, 1981.

Vol. 6: S. S. Gupta and D.-Y. Huang, Multiple Statistical Decision Theory: Recent Developments. VIII, 104 pages, 1981.

Vol. 7: M. Akahira and K. Takeuchi, Asymptotic Efficiency of Statistical Estimators. VIII, 242 pages, 1981.

Vol. 8: The First Pannonian Symposium on Mathematical Statistics. Edited by P. Révész, L. Schmetterer, and V. M. Zolotarev. VI, 308 pages, 1981.

Vol. 9: B. Jørgensen, Statistical Properties of the Generalized Inverse Gaussian Distribution. VI, 188 pages, 1981.

Vol. 10: A. A. McIntosh, Fitting Linear Models: An Application on Conjugate Gradient Algorithms. VI, 200 pages, 1982.

Vol. 11: D. F. Nicholls and B. G. Quinn, Random Coefficient Autoregressive Models: An Introduction. V, 154 pages, 1982.

Vol. 12: M. Jacobsen, Statistical Analysis of Counting Processes. VII, 226 pages, 1982.

Vol. 13: J. Pfanzagl (with the assistance of W. Wefelmeyer), Contributions to a General Asymptotic Statistical Theory. VII, 315 pages, 1982.

Vol. 14: GLIM 82: Proceedings of the International Conference on Generalised Linear Models. Edited by R. Gilchrist. V, 188 pages, 1982.

Vol. 15: K. R. W. Brewer and M. Hanif, Sampling with Unequal Probabilities. IX, 164 pages, 1983.

Vol. 16: Specifying Statistical Models: From Parametric to Non-Parametric, Using Bayesian or Non-Bayesian Approaches. Edited by J. P. Florens, M. Mouchart, J. P. Raoult, L. Simar, and A. F. M. Smith. XI, 204 pages, 1983.

Vol. 17: I. V. Basawa and D. J. Scott, Asymptotic Optimal Inference for Non-Ergodic Models. IX, 170 pages, 1983.

Vol. 18: W. Britton, Conjugate Duality and the Exponential Fourier Spectrum. V, 226 pages, 1983.

Vol. 19: L. Fernholz, von Mises Calculus For Statistical Functionals. VIII, 124 pages, 1983.

Vol. 20: Mathematical Learning Models – Theory and Algorithms: Proceedings of a Conference. Edited by U. Herkenrath, D. Kalin, W. Vogel. XIV, 226 pages, 1983.

Vol. 21: H. Tong, Threshold Models in Non-linear Time Series Analysis. X, 323 pages, 1983.

Vol. 22: S. Johansen, Functional Relations, Random Coefficients and Nonlinear Regression with Application to Kinetic Data. VIII, 126 pages. 1984.

Vol. 23: D. G. Saphire, Estimation of Victimization Prevalence Using Data from the National Crime Survey. V, 165 pages. 1984.

Vol. 24: T. S. Rao, M. M. Gabr, An Introduction to Bispectral Analysis and Bilinear Time Series Models. VIII, 280 pages, 1984.

Vol. 25: Time Series Analysis of Irregularly Observed Data. Proceedings, 1983. Edited by E. Parzen. VII, 363 pages, 1984.

ctd. on inside back cover

Lecture Notes in Statistics

Edited by D. Brillinger, S. Fienberg, J. Gani, J. Hartigan, and K. Krickeberg

33

Miklós Csörgő
Sándor Csörgő
Lajos Horváth

An Asymptotic Theory for Empirical Reliability and Concentration Processes

Springer-Verlag Berlin Heidelberg GmbH

Authors

Miklós Csörgő
Department of Mathematics and Statistics, Carleton University
Ottawa K1S 5B6, Canada

Sándor Csörgő
Lajos Horváth
Bolyai Institute, Szeged University
Aradi vértanúk tere 1, H-6720 Szeged, Hungary

Mathematics Subject Classification (1980): 62E20

ISBN 978-0-387-96359-4 ISBN 978-1-4615-6420-1 (eBook)
DOI 10.1007/978-1-4615-6420-1

Originally published by Springer-Verlag Berlin Heidelberg New York in 1986

2147/3140-543210

PREFACE

Miklós Csörgő and David M. Mason initiated their collaboration on the topics of this book while attending the CBMS-NSF Regional Conference at Texas A & M University in 1981. Independently of them, Sándor Csörgő and Lajos Horváth have begun their work on this subject at Szeged University. The idea of writing a monograph together was born when the four of us met in the Conference on Limit Theorems in Probability and Statistics, Veszprém 1982. This collaboration resulted in No. 2 of Technical Report Series of the Laboratory for Research in Statistics and Probability of Carleton University and University of Ottawa, 1983. Afterwards David M. Mason has decided to withdraw from this project. The authors wish to thank him for his contributions. In particular, he has called our attention to the reverse martingale property of the empirical process together with the associated Birnbaum-Marshall inequality (cf. the proofs of Lemmas 2.4 and 3.2) and to the Hardy inequality (cf. the proof of part (iv) of Theorem 4.1). These and several other related remarks helped us push down the moment condition to $EX^2 < \infty$ in all our weak approximation theorems.

During our work together the research of Miklós Csörgő has been supported by NSERC Canada operating grants at Carleton University. The other two authors have also been frequently supported by NSERC Canada grants of Miklós Csörgő and D.A. Dawson and by EMR Canada grants of Miklós Csörgő as Visiting Scientists in the Laboratory for Research in Statistics and Probability at Carleton University. When working at Szeged University, Miklós Csörgő has enjoyed the hospitality of the Bolyai Institute and that of Professor Károly Tandori in particular.

Our thanks go to all those people who have read and commented on the first draft of this exposition. We especially appreciate the help of Chang-Jo F. Chung (EMR Canada), John H.J. Einmahl (Catholic University, Nijmegen), Pál Révész (Technische Universität Wien), Wolfgang R. van Zwet (University of Leiden), Jon A. Wellner (University of Washington) and Brian S. Yandell (University of Wisconsin, Madison). We also express our gratitude to Mrs. Gill S. Murray of the Laboratory for Research in Statistics and Probability, Carleton University, for her expert typing of our manuscript.

CONTENTS

1. INTRODUCTION. As Barlow and Proschan (1977) write, "a unifying concept in the statistical theory of reliability and life testing is that of total time on test". The total time on test transform of a life distribution F was first discussed by Marshall and Proschan (1965) in connection with estimation problems for distributions with a monotone failure rate. The maximum likelihood estimate of F in this case is piecewise exponential (Marshall and Proschan, 1965), and the maximum likelihood estimate of the failure rate function is found (see Barlow, Bartholomew, Bremner and Brunk (1972), pp.231-242) by inverting the slopes of the least concave majorant to the total time on test process. This transform has been proved useful in various testing problems. Barlow (1968) and Barlow and Doksum (1972) studied a scale-free test of exponentiality based on the cumulative total time on test statistic which is derived from the total time on test transform. Scaled total time on test data plots were used to test exponentiality in general by Epstein and Sobel (1953) and against increasing or decreasing failure rate and other life distribution classes of interest by Barlow and Campo (1975), Klefsjö (1983a,1983b), and Doksum and Yandell (1984). Tests, based on the total time on test transform, when data are incomplete were considered by Barlow and Proschan as early as in 1969. (See also Proschan and Pyke (1967).) Barlow and Campo (1975), Barlow (1979), Chandra and Singpurwalla (1978), Langberg, León and Proschan (1980) and Klefsjö (1982) studied the geometry of the total time on test transform and characterisation results were also proved for it in these papers. For further many-sided applications see Bergman (1977a,b; 1979) and Bergman and Klefsjö (1982a,b, 1984) and their references. These papers, together with Marshall and Proschan (1972) and the corresponding sections of Barlow and Proschan (1975) provide a firm theoretical basis for the probabilistic aspects of modelling in reliability theory and life testing. Although the statistical aspects have also been touched upon in some of the above references, the investigations have mainly been centered about the exponential distribution. While this is entirely natural in view of the central role of this distribution among life distributions and, in particular, of the unique simplicity of the total time on test transform of this distribution, greater statistical flexibility is clearly required. Since questions about the exact distributions of statistics based on the total time on test transform become hopeless if we depart from exponentiality, an obstacle to such a flexibility is the lack of a general asymptotic theory for empirical total time on test processes. Although Barlow and Proschan (1977) addressed the problem, they restricted attention to the problem of pointwise convergence of total

time on test processes (see our comments in Section 7 below). Langberg, León and Proschan (1980) proved the pointwise strong consistency of the empirical total time on test function, but, apart from a result by Barlow and van Zwet (1970) to be mentioned in Section 6, we could not find a uniform Glivenko-Cantelli theorem, let alone a general weak convergence result, for total time on test processes in the literature. One of the aims of the present monograph is to construct a general convergence theory for empirical total time on test processes.

"The Lorenz curve of the distribution of 'wealth' is a graph of cumulative proportion of total 'wealth' owned, against cumulative proportion of the population owning it" as Goldie (1977) writes. He points out that Lorenz curves, and associated inequality and concentration indices, have been in use since 1905 to describe concentration and inequality in distributions of resources and in size distributions. The references in this direction of applications are too numerous to make an attempt to list them here. See, for example, Dalton (1920), Hall and Tideman (1967), Horowitz and Horowitz (1968), Bruckmann (1969), Hexter and Snow (1970), Horowitz (1970), Dasgupta, Sen and Starrett (1973), Sen (1973, 1974), Piesch (1975) and our further references in Section 16. So, again as Goldie (1977) writes, the main importance of these curves is in economics, as applied to income and wealth, and also to business concentration and the distribution of sizes of firms (Hart, 1971, 1975). As to applications outside economics, Goldie (1977) mentions bibliography (Leimkuhler, 1967), the distribution of scientific grants (House of Commons (1975), Allison et al. (1976)), fishery (Thompson, 1976), and politics (Alker, 1965). Wold (1935), Gastwirth (1972), Kakwani and Podder (1973) and, unaware of Goldie (1977), Sendler (1979) considered the problem of estimating the theoretical Lorenz curve from data, while Gastwirth (1971, 1972), Chandra and Singpurwalla (1978) studied various theoretical properties of the Lorenz curve and the associated Gini index. Gail and Gastwirth (1978a,b) proposed scale-free tests for exponentiality based on the Lorenz curve and the Gini statistic. Chandra and Singpurwalla (1978) stated a weak convergence result for empirical Lorenz processes, although they only proved pointwise convergence (see our remarks in Section 12 below). It was Goldie (1977) who provided a remarkable and thorough convergence theory for empirical Lorenz and, what he calls, concentration processes. The latter processes are in fact inverse Lorenz processes whose potential usefulness in econometrics suggests that they are at least as important as Lorenz processes themselves.

Chandra and Singpurwalla (1978, 1981) seem to be the first who pointed out the important observation that there is a close relationship between the total time on test transform and the theoretical Lorenz curve, and, in particular, between the various indices associated with these transforms, such as the cumulative total time on test and the Gini index. However, no connecting theories have so far been constructed for the two kinds of processes in question.

The primary aim of the present monograph is to build up a unified asymptotic theory for empirical total time on test, Lorenz, and concentration processes. The feasibility of such a unified theory was outlined by M. Csörgő (1983) under some unnecessarily strong conditions. Rather than using results directly from the theory of general quantile processes, here we work out special techniques tailored exactly for the present reliability and economic processes. In the course of our work it became clear, too, that there are other means for measuring inequality or diversity, and concentration, different from the usual Lorenz curve. Accordingly, new Lorenz type empirical processes are introduced, for instance, the empirical Shannon and the associated empirical redundancy processes. Similarly modified total time on test processes are also considered. All these processes will fit into our unified theory.

One of the simplest common ingredients of the processes considered is closely related to mean residual life processes. They are important in biometry, and have been considered, for example, by Wilson (1938), Chiang (1960, 1968), Bryson and Siddiqui (1969), Gross and Clark (1975), and Hollander and Proschan (1975) from the statistical point of view, while their probabilistic aspects were nicely summarised by Hall and Wellner (1981). The statistical theory for the convergence of mean residual life processes culminates in Yang (1978), and Hall and Wellner (1979). Our unified theory covers the just mentioned weak convergence results, while providing also further insights into the nature of mean residual life processes.

A look at Goldie's (1977) paper, one of the strongest and most difficult mathematical papers in the classical weak convergence theory in C and various Skorohod spaces, shows that the processes in question are far from being easy probabilistic objects. Although possible in principle, it would be very hard to achieve our aimed at unified theory by the traditional two-part pattern for proofs of invariance principles (finite-dimensional distributions, tightness).

Our approach is the approximation method, which also makes it possible to see clearly what additional assumptions are needed on the underlying distribution in order that the considered empirical processes be similar to corresponding Gaussian processes in a stronger sense than the weak invariance principle obtained in the "first step" of this approach. The second "strong step" then provides loglog law consistency rates (uniformly) and makes possible the transition of the fluctuational behaviour of the approximating Gaussian processes over to the empirical processes up to the rate of approximation, depending again on the regularity of the underlying distribution. This monograph is therefore also methodological, demonstrating the strength of an approach in a non-trivial situation.

The basic approximation theory for the ordinary empirical and quantile processes, and for some of their transforms, is contained in the books of M. Csörgő and Révész (1981), and M. Csörgő (1983). All the results we need here are described in detail in our Section 2. However, this preliminary Section 2 also contains a number of new results which are of interest on their own right.

We now turn this introduction towards more technical terms. In order to obtain our unified theory we assume throughout that the underlying nondegenerate life distribution function F, $F(0) = 0$, is <u>continuous</u>. Hence our results will not cover those of Goldie (1977) and Sandler (1982) for the weak convergence of Lorenz processes, and those of Goldie (1977) for his concentration processes in Skorohod's M_1 topology. Presently however, we do not assume the continuity of the quantile function

$$Q(y) = F^{-1}(y), \quad 0 \le y \le 1. \tag{1.1}$$

Here and throughout, for a nondecreasing right-continuous function $k(x)$ on the line, the right-continuous inverse k^{-1} is defined as

$$k^{-1}(y) = \inf\{x : k(x) > y\}. \tag{1.2}$$

To avoid confusion, a function, deterministic or random, without discontinuities of the second kind will always be defined to be right-continuous. Throughout $F(0) = 0$ will be assumed, except when the contrary is explicitly stated in Point (6) of Section 8 and in Section 9.

Let $X_1,\ldots,X_n$ be independent random variables with common distribution function F, a random sample of n observations on X. Introduce

$$W_{k:n} = (n+1-k)(X_{k:n}-X_{k-1:n}),\ k=1,\ldots,n,$$

with $X_{0:n} \equiv 0$, where $X_{1:n} \le \ldots \le X_{n:n}$ is the ordered sample. According to Barlow and Proschan (1975), p. 61, or Langberg, León and Proschan (1980), the <u>total time on test</u> up to the kth order statistic, $T(X_{k:n})$, is defined by $T(X_{k:n}) = \Sigma_{i=1}^{k} W_{i:n}$ for $k=1,\ldots,n$. If we assume that n items are placed on test at time 0 and that the successive failures are obtained at times $X_{1:n} \leq \ldots \leq X_{n:n}$, then $W_{k:n}$ represents the total test time observed between $X_{k-1:n}$ and $X_{k:n}$, and $T(X_{k:n})$ represents the total test time observed between 0 and $X_{k:n}$. We define the nth total time on test function as

$$\text{(1.3)} \qquad H_n^{-1}(u) = \frac{1}{n} T(X_{[nu]+1:n})$$

$$= \frac{1}{n} \Sigma_{i=1}^{[nu]+1} W_{i:n}$$

$$= \frac{1}{n} \Sigma_{i=1}^{[nu]} X_{i:n} + (1 - \frac{[nu]}{n}) X_{[nu]+1:n}$$

for $0 \leq u < 1$ and

$$\text{(1.4)} \qquad H_n^{-1}(1) = \lim_{u \uparrow 1} H_n^{-1}(u) = \frac{1}{n} \Sigma_{i=1}^{n} X_i = \bar{X}_n ,$$

where $[\cdot]$ is the integer part function. This is a nondecreasing right-continuous random function, and we define its theoretical counterpart, the total time on test transform of F, as

$$\text{(1.5)} \qquad H_F^{-1}(u) = \int_0^u (1-y)\,dQ(y) + t_F$$

$$= (1-u)Q(u) + \int_0^u Q(y)\,dy, \quad 0 \leq u \leq 1,$$

where

$$t_F = \sup\{t : F(t) = 0\}$$

is the lower endpoint of the support of F. Although we assumed $F(0)= 0$, that is, $t_F \geq 0$, we note that this definition of H_F^{-1} is meaningful for all cases whenever $t_F > -\infty$. In fact, in Section 9 we shall definitely allow $t_F < 0$. Above and throughout we use the convention

$$\int_a^b = \int_{[a,b)} ,$$

$a < b$, for all occurring Lebesgue-Stieltjes integrals. Assuming that, together with F, the quantile function $Q = F^{-1}$ is also continuous (on $[0,1)$), and this assumption will always be stipulated whenever we talk about total time on test processes, we have

$$\text{(1.6)} \qquad H_F^{-1}(u) - t_F = \int_0^{Q(u)} (1-F(y))\,dy, \quad 0 \leq u \leq 1,$$

agreeing with its usual definition. Clearly $H_F^{-1}(u) \leq H_F^{-1}(1)$ for all $u \in [0,1)$, and hence it is a finite function on the whole interval $[0,1]$ if and only if

(1.7) $$\mu = EX = H_F^{-1} - t_F < \infty .$$

Here, and also in the sequel, X denotes a generic random variable with distribution function F. In the literature on life distributions it is quite naturally assumed that $t_F = 0$. In general little is gained by allowing $t_F \neq 0$, $t_F > -\infty$, but for the sake of certain problems in Sections 8 and 9 we nevertheless allow this possibility. So all the statements in the first eight sections are meant with the above general definitions of H_F^{-1} in (1.5) or (1.6). On the other hand, the possibility of $t_F \neq 0$ requires only trivial extra considerations, and hence, in the proofs of all the statements in the first eight sections we assume $t_F = 0$ without loss of generality.

Upon introducing the Lorenz curve of F as

(1.8) $$L_F(u) = \frac{1}{\mu} \int_0^u Q(y)\,dy, \quad 0 \le u \le 1,$$

we have the relation

(1.9) $$H_F^{-1}(u) - t_F = (1-u)Q(u) + \mu L_F(u), \quad 0 \le u \le 1.$$

In econometrics, $L_F(u)$ is commonly interpreted as the fraction of total income that the holders of the lowest u^{th} fraction of incomes possess. For our unified theory it will be convenient to define the empirical Lorenz curve as

(1.10) $$L_n(u) = \begin{cases} \frac{1}{\overline{X}_n} \frac{1}{n} \sum_{i=1}^{[nu]+1} X_{i:n}, & 0 \le u < 1, \\ 1, & u = 1, \end{cases}$$

in order to enable ourselves to tie it up with $H_n^{-1}(u)$. Here $\overline{X}_n = n^{-1}\Sigma_{i=1}^n X_i$ is the sample mean. This definition, at least with $[nu]$ instead of our $[nu]+1$, was used by Taguchi (1968).

Goldie's (1977) definition of the empirical Lorenz curve is different from ours and agrees with that of Gastwirth (1971, 1972). It is the polygonal line joining the points

$$\left(\frac{k}{n}, \frac{\sum_{j=0}^{k} X_{j:n}}{\sum_{j=1}^{n} X_{j:n}} \right) \quad (k=0,1,\ldots,n) \text{ with } X_{0:n} \equiv 0.$$

Denoting this curve by $L_n^*(\cdot)$, we have

$$\Delta_n^* = \sup_{0 \le u \le 1} |L_n(u) - L_n^*(u)| = \frac{X_{n:n}}{\sum_{k=1}^{n} X_{k:n}}$$

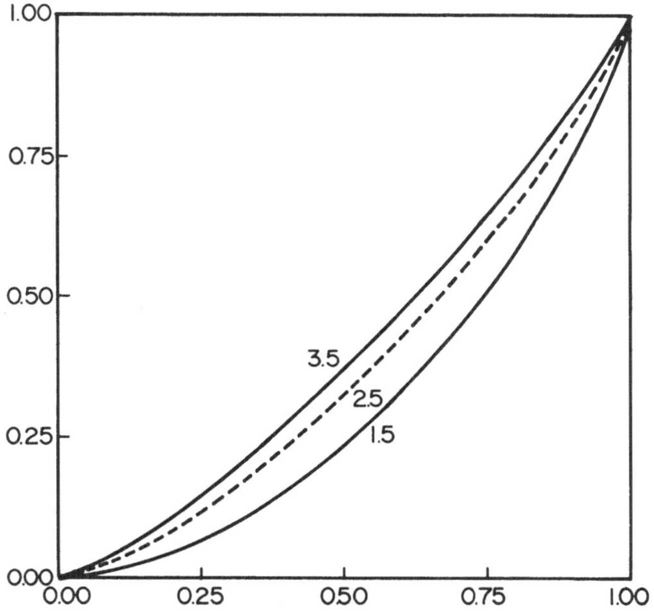

Figure 1.a

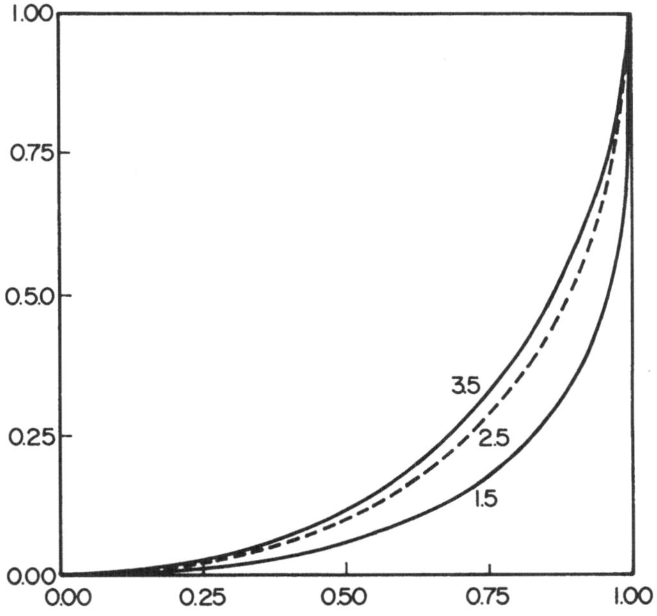

Figure 1.b

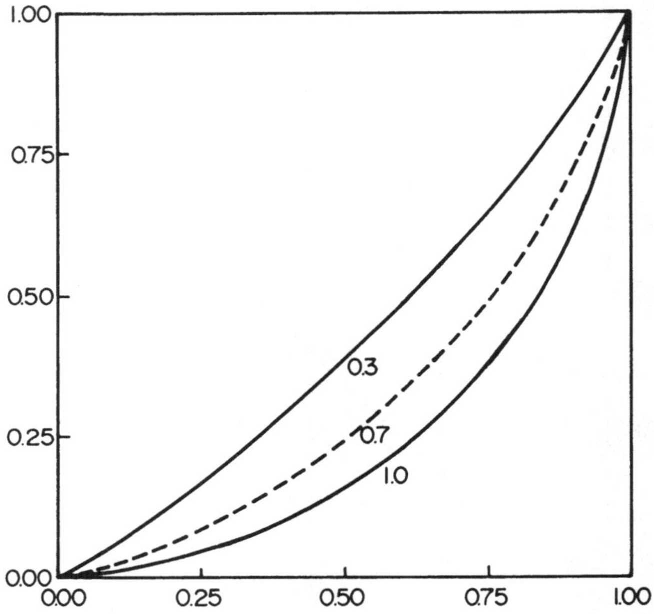

Figure 1.c

Figure 1.a : Lorenz curve of $F(x) = 1-e^{-x^c}$, $x \geq 0$, with $c = 1.5$, 2.5 and 3.5

Figure 1.b : Lorenz curve of $F(x) = 1-(1+x)^{-c}$, $x \geq 0$, with $c = 1.5$, 2.5 and 3.5

Figure 1.c : Lorenz curve of $F(x) = \Phi(\frac{\log x}{c})$, $x > 0$, with $c = 0.3$, 0.7 and 1 .

and this converges to zero almost surely as $n \to \infty$, as Goldie (1977) notes, if $\mu < \infty$, since then Dugué's (1958, p.71) theorem ensures that $X_{n:n}/n \to 0$ a.s. In fact, by the same reason and the central limit theorem we also have that $n^{\frac{1}{2}}\Delta_n^* \to 0$ in probability, provided that $EX^2 < \infty$. Goldie (1977) also deals with a modified empirical Lorenz curve

$$\tilde{L}_n(u) = \begin{cases} \frac{1}{\bar{X}_n} \int_0^{F_n^{-1}(u)} x dF_n(x), & 0 \leq u < 1, \\ 1 & , \quad u = 1, \end{cases}$$

where F_n^{-1} is the inverse to the right-continuous empirical distribution function F_n of $X_1, \ldots, X_n$, i.e.,

$$F_n(x) = \frac{1}{n} \#\{1 \leq i \leq n: X_i \leq x\} .$$

He shows that $\tilde{L}_n \to \tilde{L}_F$ almost surely in the Skorohod J_1 topology, where

$$\tilde{L}_F(u) = \begin{cases} \frac{1}{\mu} \int_0^{Q(u)} x dF(x) & , \quad 0 \leq u < 1, \\ \tilde{L}_F(1-) = 1 & , \quad u = 1, \end{cases}$$

without the here assumed continuity of F even. This modified theoretical Lorenz curve $\tilde{L}_F$ does not generally coincide with L_F when F may have discontinuities. They are the same, however, when F is continuous, and this is assumed here throughout. Then, of course, Skorohod's J_1 convergence may be replaced by uniform convergence, and thus the three empirical curves L_n, L_n^* and $\tilde{L}_n$ are asymptotically equivalent.

In the second section we fix the basic setting for our approximation results and list the appropriate results for ordinary empirical and quantile processes, together with those for their Gaussian counterparts, we require later. As we have already noted, some of these results are new. With a familiarity of the notation from Section 2 ($E_n(\cdot) = F_n(Q(\cdot))$ and $U_n(\cdot)$ standing, respectively, for the uniform empirical distribution function and quantile function of the transformed sample $F(X_1), \ldots, F(X_n)$) it is advantageous for a possible reader at this stage to have a quick glance at the easy but basic representations in (6.1) and (10.1) for the empirical total time on test function H_n^{-1} and for the unscaled empirical Lorenz curve $G_n = \bar{X}_n L_n$, respectively. These integral representations at once suggest that two kinds of common ingredients will be met when dealing with total time on test and Lorenz processes. These are integrals of ordinary empirical processes and empirical increments of certain Brownian bridge sequences. These two kinds of auxiliary processes are investigated in Sections 3 and 5, respectively. Then Section 6 is devoted to the study of the convergence problems of the total time on test empirical process

(1.11) $$t_n(u) = n^{\frac{1}{2}}(H_n^{-1}(u) - H_F^{-1}(u)), \quad 0 \leq u \leq 1,$$

while Section 7 to those of its scaled version

(1.12) $$s_n(u) = n^{\frac{1}{2}}(D_n^{-1}(u)-D_F^{-1}(u)), \quad 0 \leq u \leq 1,$$

where

$$D_F^{-1}(u) = \frac{1}{\mu} H_F^{-1}(u)$$

and

$$D_n^{-1}(u) = \frac{1}{\bar{X}_n} H_n^{-1}(u) .$$

It was convenient to unscale the empirical Lorenz process

(1.13) $$\ell_n(u) = n^{\frac{1}{2}}(L_n(u)-L_F(u), \quad 0 \leq u \leq 1,$$

and to deal with the unscaled version first in Section 10, implying at once the corresponding results for the original $\ell_n(\cdot)$ in Section 11. The structure of all the so far mentioned sections, beginning with the third one, follows the same pattern: strong uniform consistency, weak approximation (implying always a weak convergence result in the supremum norm), and strong approximation (implying a uniform law of the iterated logarithm and other strong laws for the size of the increments of the corresponding empirical processes). Their method of proof also illuminates the essence of the stochastic nature of these processes. The results obtained, and their conditions together with various corollaries and the structure of the limiting processes, are discussed in Section 8 for total time on test processes and in Section 12 for Lorenz processes. The said discussion in Section 8 leads to some new problems in connection with scale and shift families, and to corresponding modifications of total time on test processes and to their two-sided analogues. Section 9 contains these corresponding considerations.

The convergence theory for the important concentration process

(1.14) $$c_n(u) = n^{\frac{1}{2}}(L_n^{-1}(u)-L_F^{-1}(u)), \quad 0 \leq u \leq 1,$$

of Goldie is worked out in Section 13 and the there obtained triad of results is discussed in Section 14. Because of the complex nature of this inverse process $c_n(\cdot)$, a much more complicated quantile-type process than the usual quantile process, our Section 13 is perhaps the

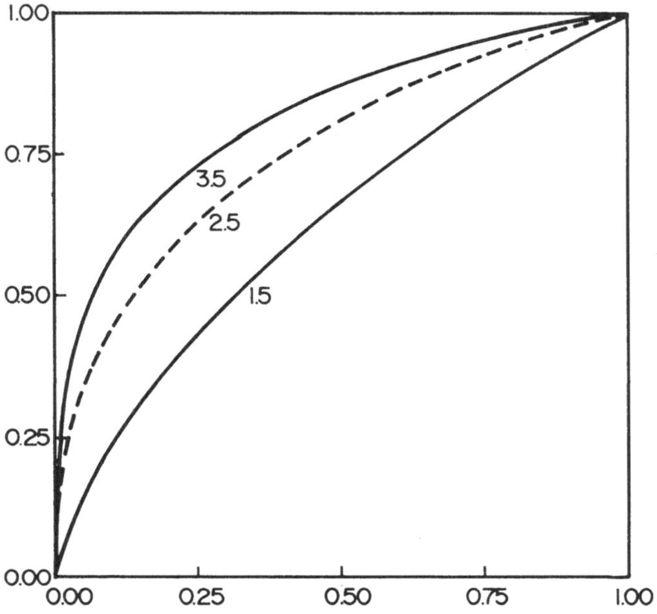

Figure 2.a

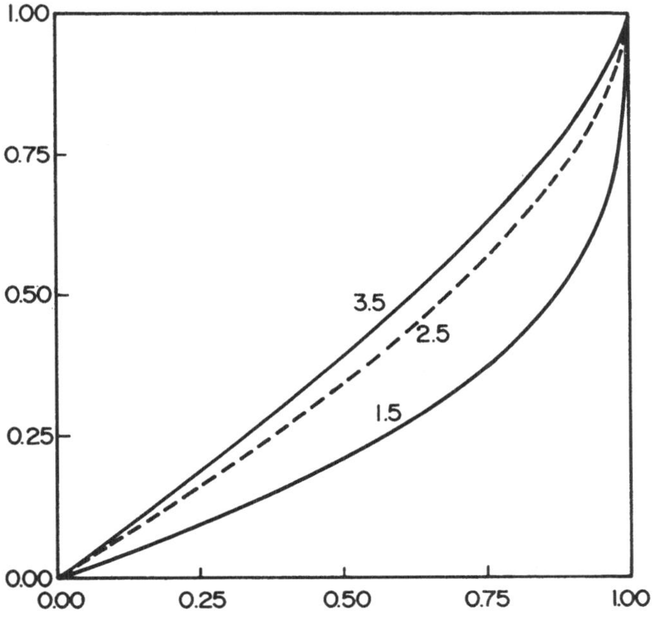

Figure 2.b

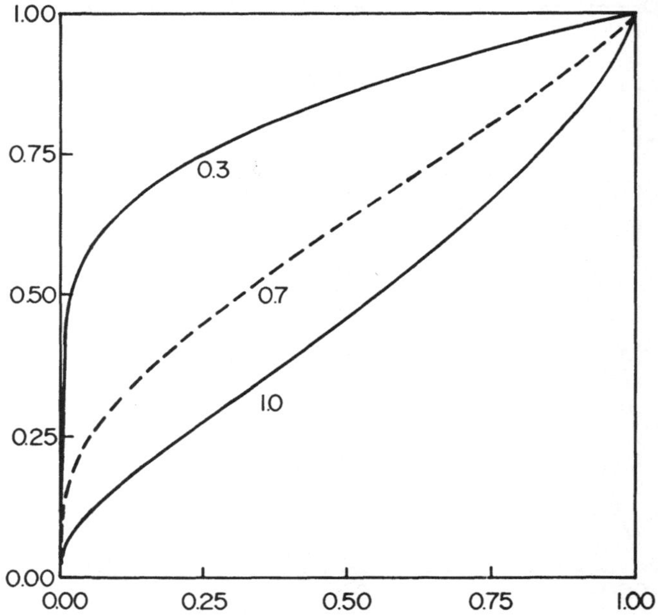

Figure 2.c

Figure 2.a : Scaled total time on test of $F(x) = 1-e^{-x^{-c}}$, $x \geq 0$, with $c = 1.5$, 2.5 and 3.5

Figure 2.b : Scaled total time on test of $F(x) = 1-(1+x)^{-c}$, $x \geq 0$, with $c = 1.5$, 2.5 and 3.5

Figure 2.c : Scaled total time on test of $F(x) = \Phi(\frac{\log x}{c})$, $x > 0$, with $c = 0.3$, 0.7 and 1.

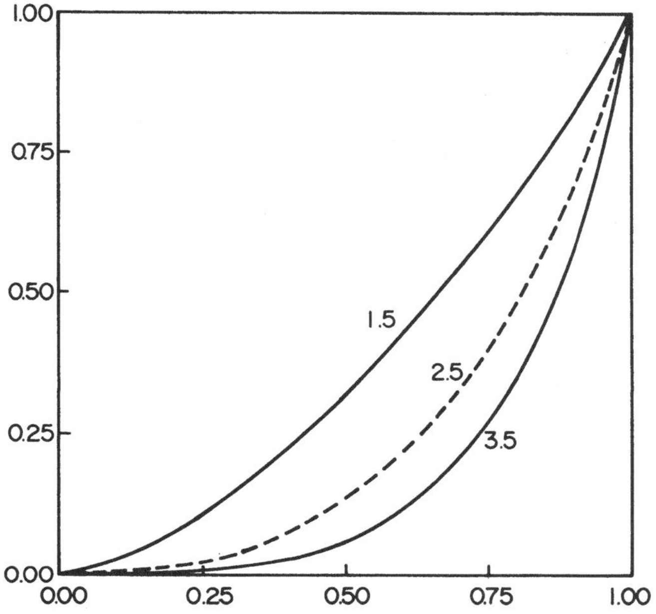

Figure 3.a

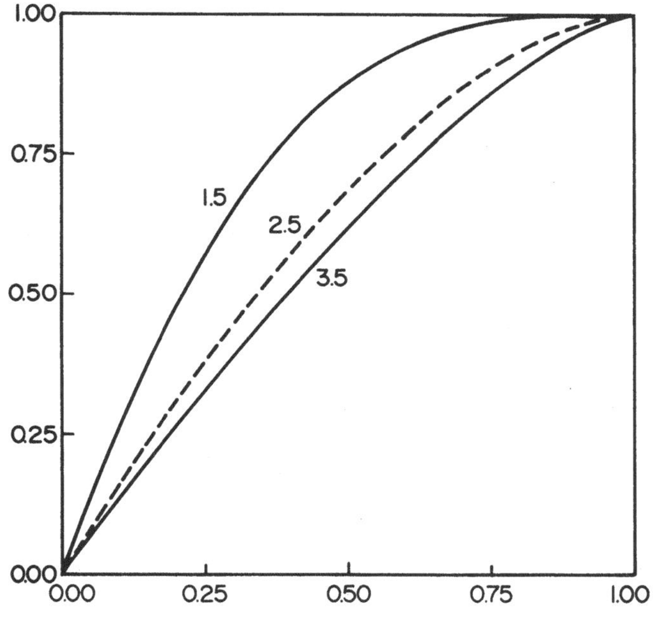

Figure 3.b

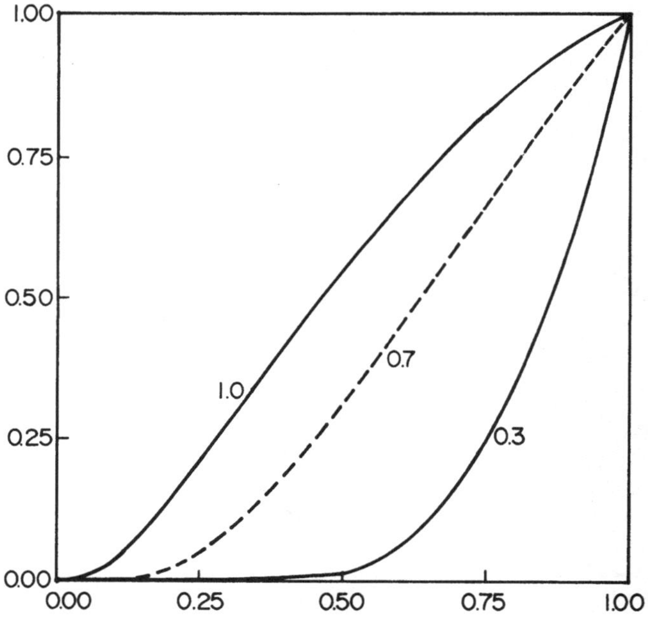

Figure 3.c

Figure 3.a : D_F , inverse of the scaled total time on test, $F(x) = 1-e^{-x^{-c}}$, $x \geq 0$, with $c = 1.5$, 2.5 and 3.5

Figure 3.b : D_F , inverse of the scaled total time on test, $F(x) = 1-(1+x)^{-c}$, $x \geq 0$, with $c = 1.5$, 2.5 and 3.5

Figure 3.c : D_F , inverse of the scaled total time on test, $F(x) = \Phi(\frac{\log x}{c})$, $x > 0$, with $c = 0.3$, 0.7 and 1.

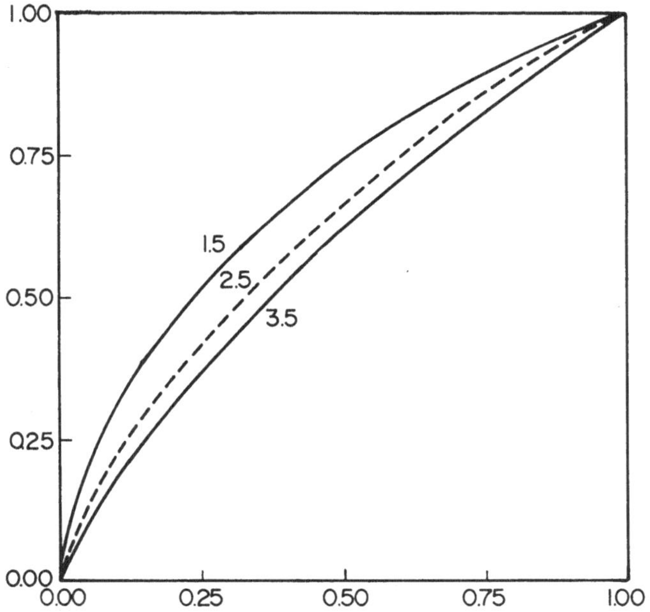

Figure 4.a

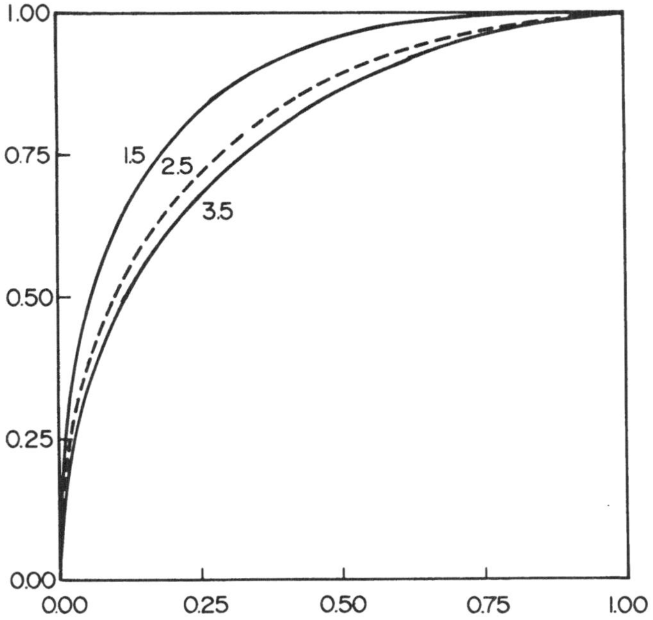

Figure 4.b

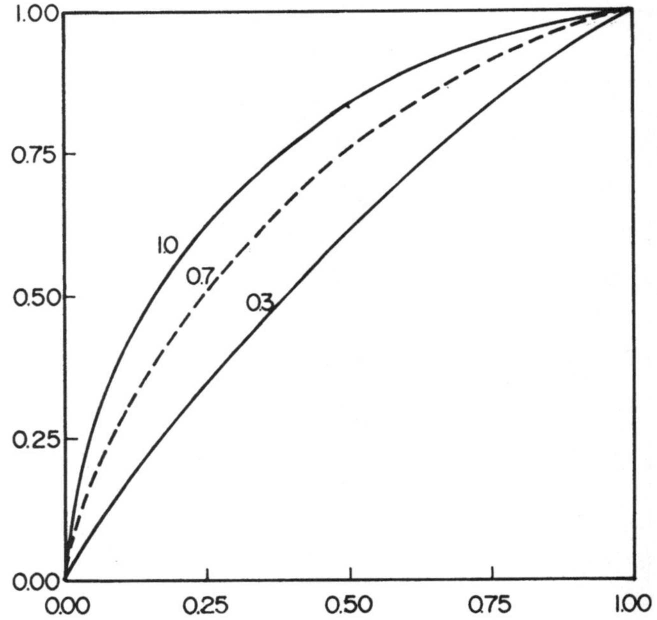

Figure 4.c

Figure 4.a : Goldie concentration curve of $F(x) = 1-e^{-x^{-c}}$, $x \geq 0$, with $c = 1.5, 2.5$ and 3.5

Figure 4.b : Goldie concentration curve of $F(x) = 1-(1+x)^{-c}$, $x \geq 0$, with $c = 1.5, 2.5$ and 3.5

Figure 4.c : Goldie concentration curve of $F(x) = \Phi(\frac{\log x}{c})$, $x > 0$, with $c = 0.3, 0.7$ and 1.

most difficult technically. Many elements of the approximation technique, already available for the preceding processes from Section 2, but not fitting the concentration process, had to be worked out separately here.

New Lorenz type and concentration processes are introduced and investigated in Section 15. While they are formally more general or complicated than the Lorenz process and the Goldie concentration process, the strong uniform consistency, and weak and strong approximation results for them easily follow from those for the ordinary Lorenz process and its inverse.

The first and easiest kind of auxiliary processes, the integrals of ordinary empirical processes treated in Section 3, are closely related to the mean residual life process

(1.15) $$z_n(x) = n^{\frac{1}{2}}(M_n(x) - M_F(x)), \quad 0 \leq x < \infty,$$

where

$$M_F(x) = E(X - x \mid X > x)$$

$$= \frac{1}{1-F(x)} \int_x^\infty (1-F(t))\,dt$$

is the mean residual life function at age x. The empirical counterpart of M_F is

$$M_n(x) = M_{F_n}(x) = \frac{1}{1-F_n(x)} \int_x^\infty (1-F_n(t))\,dt.$$

Our usual triad of results for z_n is derived directly in Section 4 from that in Section 3, and a few consequences of the strong approximation result are also discussed there.

To the best of our knowledge the regularity conditions for all of our weak approximation results are always weaker than the regularity conditions for the corresponding weak convergence results existing in the literature, apart from the continuity of F when speaking about the mean residual life process and the continuity of Q when speaking about the Lorenz and the Goldie concentration processes. No strong versions of these results are available in the literature. We are not aware of any similar results for total time on test processes.

Section 16 contains easy consequences on the strong consistency and asymptotic normality for a number of functionals of the considered processes, widely used in the applied literature as measures, or indices of inequality, diversity, and concentration.

Finally, in Section 17 we introduce the notion of bootstrapping empirical functionals. We discuss in detail the practical applications

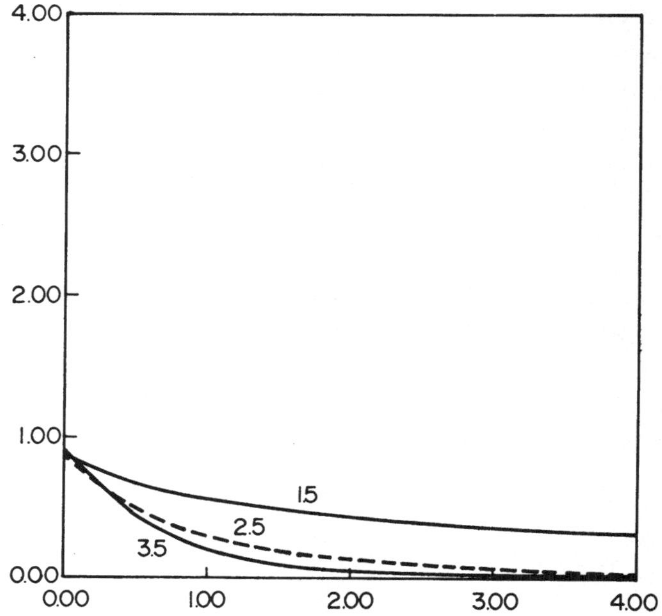

Figure 5.a

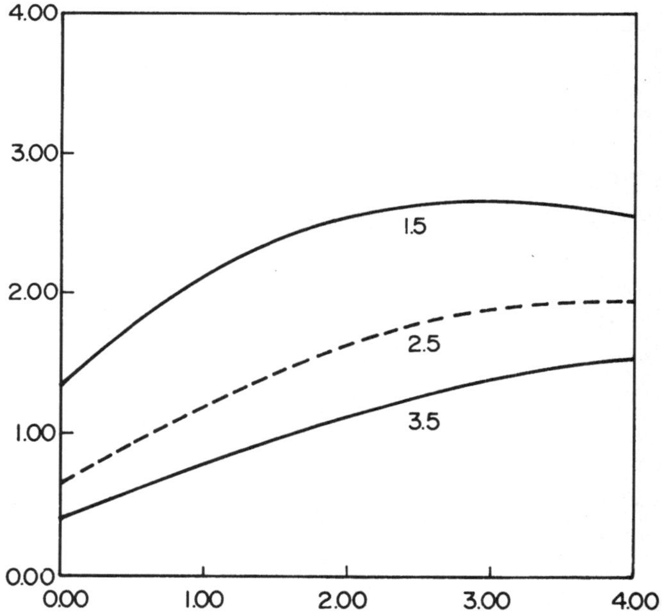

Figure 5.b

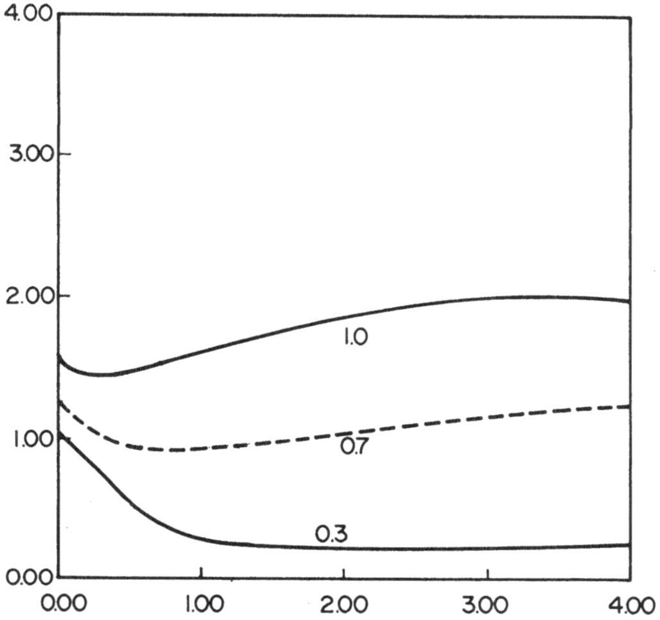

Figure 5.c

Figure 5.a : Mean residual life function of $F(x) = 1-e^{-x^{-c}}$, $x \geq 0$, with $c = 1.5, 2.5$ and 3.5

Figure 5.b : Mean residual life function of $F(x) = 1-(1+x)^{-c}$, $x > 0$, with $c = 1.5, 2.5$ and 3.5

Figure 5.c : Mean residual life function of $F(x) = \Phi(\frac{\log x}{c})$, $x > 0$, with $c = 0.3, 0.7$ and 1.

of the bootstrap, and establish also the necessary mathematical tools.

Unless otherwise specified, all convergence and rate of convergence statements will be meant as $n \to \infty$.

2. THE BASIC SETTING FOR THE APPROXIMATIONS AND VARIOUS PRELIMINARIES.

Without loss of generality we assume that our basic sequence $X_1, X_2, \ldots$ is defined on an appropriate probability space (Ω, A, P) such that for the resulting uniform empirical process

$$\alpha_n(y) = n^{\frac{1}{2}}(y - E_n(y)), \quad 0 \le y \le 1,$$

of $U_1 = F(X_1), \ldots, U_n = F(X_n)$, the approximation

$$\sup_{0 \le y \le 1} |\alpha_n(y) - n^{-\frac{1}{2}} K(y,n)| \overset{a.s.}{=} O((\log n)^2 / n^{\frac{1}{2}}) \tag{2.1}$$

of Komlós, Major and Tusnády (1975) holds true (see also Theorem 4.4.3 in M. Csörgő and Révész (1981)). Here E_n is the (right-continuous) empiric distribution function of $U_1, \ldots, U_n$, i.e., $E_n(y) = F_n(Q(u))$ for all $y \in [0,1]$, and $K(y,x)$, $0 \le y \le 1$, $0 \le x < \infty$, is a Kiefer process, that is, a separable real valued mean zero two-parameter Gaussian process, with covariance function

$$EK(y,x)K(u,t) = \min(x,t)(\min(y,u) - yu)$$

for any $u, y \in [0,1]$ and $t, x \in [0,\infty)$. Whenever we write

$$R_n \overset{a.s.}{=} O(r_n)$$

for a sequence of random variables R_n and positive constants r_n, we mean that

$$\limsup_{n \to \infty} R_n / r_n \le C \quad \text{a.s.}$$

with a non-random positive constant C. We introduce now the quantile function of $U_1, \ldots, U_n$ as

$$U_n(y) = \begin{cases} U_{k:n}, & \frac{k-1}{n} \le y < \frac{k}{n}, \quad k = 1, \ldots, n, \\ U_{n:n}, & y = 1, \end{cases}$$

where $U_{1:n} \le \ldots \le U_{n:n}$ is the ordered sample $U_1, \ldots, U_n$. By our general inverse definition in (1.2) we have

$$U_n(y) = E_n^{-1}(y), \quad 0 \le y \le 1, \tag{2.2}$$

and, according to Theorem 4.5.3 and Remark 4.5.1 of M. Csörgő and Révész (1981), for the uniform quantile process

$$u_n(y) = n^{\frac{1}{2}}(U_n(y) - y), \quad 0 \le y \le 1,$$

we have

$$\sup_{0 \le y \le 1} |u_n(y) - n^{-\frac{1}{2}} K(y,n)| \overset{a.s.}{=} O(n^{-\frac{1}{4}} (\log\log n)^{\frac{1}{4}} (\log n)^{\frac{1}{2}}) | \tag{2.3}$$

with the same Kiefer process as in (2.1). The constant of the latter $O(\cdot)$ rate is $2^{-1/4}$, as in Kiefer (1970).

Let $q(y) > 0$ be a continuous function on $(0,1)$ which is non-decreasing on $[0,1/2]$ and is symmetric about $y = 1/2$. For later references we collect these functions into the set

(2.4) $Q^* = \{q: q > 0$ on $(0,1)$ is continuous, $\nearrow$ on $[0,1/2]$, and symmetric about $1/2\}$.

A function $q \in Q^*$ will be called an O'Reilly weight function if and only if

$$\int_0^{1/2} \frac{1}{t} \exp\left(-\varepsilon \frac{q^2(t)}{t}\right) dt < \infty \quad \text{for all} \quad \varepsilon > 0. \tag{2.5}$$

It will be more convenient to discuss this condition following Lemma 2.4. Now we formulate an easy result which, on the other hand, will be a very important technical tool throughout. It says that the weak approximation versions of O'Reilly's weak convergence results for the uniform empirical and quantile processes hold simultaneously with the same sequence

$$B_n(y) = n^{-1/2} K(y,n), \quad 0 \le y \le 1, \ n=1,2,\ldots \tag{2.6}$$

of approximating Brownian bridges figuring in (2.1) and (2.3). This settles a problem posed by Shorack (1979), although he probably had in mind the derivation of this result directly from (2.1) and (2.3) without using O'Reilly's proof as we do here. In this respect see also the discussion following Lemma 2.4 referred to above.

LEMMA 2.1. *If q is an O'Reilly weight function then*

$$\sup_{0\le y\le 1} \left|\frac{\alpha_n(y)-B_n(y)}{q(y)}\right| + \sup_{\frac{1}{n}\le y\le \frac{n-1}{n}} \left|\frac{u_n(y)-B_n(y)}{q(y)}\right| \xrightarrow{P} 0. \tag{2.7}$$

Proof. Letting $0 < \theta < 1/2$ and noting that $1/q$ is bounded on $[\theta, 1-\theta]$ for any such θ, it follows from (2.1) and (2.3) that

$$\sup_{\theta\le y\le 1-\theta} \left|\frac{\alpha_n(y)-B_n(y)}{q(y)}\right| + \sup_{\theta\le y\le 1-\theta} \left|\frac{u_n(y)-B_n(y)}{q(y)}\right| \xrightarrow{a.s.} 0.$$

The result then follows by showing

$$\lim_{\theta\to 0} \limsup_{n\to\infty} P\{|\alpha_n(y)| > \varepsilon q(y) \text{ for some } 0 < y \le \theta\} = 0 \tag{2.*}$$

and

$$\lim_{\theta\to 0} \limsup_{n\to\infty} P\{|u_n(y)| > \varepsilon q(y) \text{ for some } \tfrac{1}{n} < y \le \theta\} = 0 \tag{2.**}$$

for all $\varepsilon > 0$, and that the same holds with B_n replacing α_n, where, of course, the corresponding probability for B_n does not depend on n. These relations are the ones whose proofs constitute the sufficiency parts of O'Reilly's (1974) theorems (cf. his Proposition 2.1, and relations (3.4) and (4.33)). Condition (2.5) is also necessary for the first term of (2.7) to go to zero in probability (cf. Theorem 2 of O'Reilly (1974)).

We note that the sup in the quantile term of (2.7) cannot be extended to (0,1) if $q(0) = 0$. (See M. Csörgő, S. Csörgő, Horváth and Révész (1983).) A trivial but interesting corollary to Lemma 2.1 is that

$$\sup_{\frac{1}{n} \leq y \leq \frac{n-1}{n}} \left| \frac{\alpha_n(y) - u_n(y)}{q(y)} \right| \xrightarrow{P} 0.$$

We shall repeatedly apply James' (1975) law of the iterated logarithm for weighted empirical processes. In fact, disregarding logarithmic factors, the following simple consequence of James' law will suffice for our purposes:

$$(2.8) \qquad \sup_{0<y<1} \frac{|\alpha_n(y)|}{(y(1-y))^{\frac{1}{2}-\delta}} \stackrel{a.s.}{=} O((\log\log n)^{\frac{1}{2}}) \quad \text{for any } \delta > 0.$$

We shall also require a counterpart of (2.8) for the approximating Brownian bridges of (2.6) generated by the single Kiefer process K. The following lemma is interesting in that it is not true, according to James' (1975) full law, when B_n is replaced by α_n.

LEMMA 2.2. <u>For</u> B_n <u>in</u> (2.6) <u>and</u> $h(y) = (y(1-y)\log\log \frac{1}{y(1-y)})^{\frac{1}{2}}$ <u>we have</u>

$$\limsup_{n\to\infty} \frac{1}{(\log\log n)^{\frac{1}{2}}} \sup_{0<y<1} \frac{|B_n(y)|}{h(y)} \leq 2 \quad \text{a.s.}$$

<u>Proof</u>. The exact loglog law of Corollary 1.15.2 in M. Csörgő and Révész (1981) states that

$$\limsup_{n\to\infty} \sup_{0<y<1} \frac{|B_n(y)|}{\{y(1-y)\log\log n/(y(1-y))\}^{\frac{1}{2}}} \stackrel{a.s.}{=} 2.$$

The lemma then follows by showing that

$$((\log a)(\log b))^{-1} \leq (\log (a+b))^{-1}$$

for any $b \in [\log 4, \infty)$ if a is large enough, independently of b, and by substituting $a = \log n$ and $b = \log (y(1-y))^{-1}$.

The lemma of course implies the parallel of (2.8):

(2.9) $$\limsup_{n\to\infty} \frac{1}{(\log\log n)^{\frac{1}{2}}} \sup_{0<y<1} \frac{|B_n(y)|}{(y(1-y))^{\frac{1}{2}-\delta}} \leq 2, \quad \delta > 0.$$

A famous result of Kiefer (1970) (Theorem 5.2.1 in M. Csörgő and Révész (1981); cf. also (6.2.7), (6.2.11) in M. Csörgő (1983), and see Shorack (1982) for a new derivation of it by the approximation method) implies that the simultaneous approximation in (2.1) and (2.3) cannot be done any better. However, in Section 6 we shall require a result for the uniform quantile process which follows from a better rate result than that in (2.3). This can be done, separately, by a necessarily different sequence $\{\tilde{B}_n(y),\ 0 \leq y \leq 1\}$ of Brownian bridges. Extending our basic space (Ω, A, P) if necessary (cf. Lemma 3.1.2 in M. Csörgő (1983)), Theorem 4.5.2 of M. Csörgő and Révész (1981) implies that

(2.10) $$\sup_{0\leq y\leq 1} |u_n(y) - \tilde{B}_n(y)| \overset{a.s.}{=} O((\log n)n^{-\frac{1}{2}}).$$

Note that it is not known whether this sequence $\{\tilde{B}_n\}$ can be induced by one single Kiefer process, as in (2.6), or not.

Now for any Brownian bridge B on (Ω, A, P) we have with a positive constant C_o that

(2.11) $$P\{\sup_{0\leq y\leq \varepsilon} |B(y)| > C_o\varepsilon^{\frac{1}{2}}(\log n)^{\frac{1}{2}}\} \leq \frac{1}{n^2} .$$

This follows from our exact knowledge of the distribution of $\sup\{|W(y)| : 0 \leq y \leq \varepsilon\}$ and the equality $B(y) = W(y)-yW(1)$, $0\leq y\leq 1$, in distribution, where W is a Wiener process.

Relations (2.10) and (2.11) now imply that, for any sequence $\varepsilon_n \to 0$, we have

(2.12) $$\sup_{0\leq y\leq \varepsilon_n} |u_n(y)| \overset{a.s.}{=} O(\max(n^{-\frac{1}{2}}\log n,\ \varepsilon_n^{\frac{1}{2}}(\log n)^{\frac{1}{2}})),$$

and since the inequality (2.11) also holds when the domain of the sup is $[1-\varepsilon, 1]$, we also have

(2.13) $$\sup_{1-\varepsilon_n\leq y\leq 1} |u_n(y)| \overset{a.s.}{=} O(\max(n^{-\frac{1}{2}}\log n,\ \varepsilon_n^{\frac{1}{2}}(\log n)^{\frac{1}{2}})).$$

Of course, on the whole interval (2.3) implies

(2.14) $$\limsup_{n\to\infty} (\log\log n)^{-\frac{1}{2}} \sup_{0\leq y\leq 1} |u_n(y)| \overset{a.s.}{=} 1/\sqrt{2} ,$$

upon applying the same law (Corollary 1.15.1 in M. Csörgő and Révész (1981)) for the Kiefer process, namely that

(2.15) $$\limsup_{n\to\infty} (\log\log n)^{-\frac{1}{2}} \sup_{0\le y\le 1} |B_n(y)| \overset{a.s.}{=} 1/\sqrt{2} .$$

We turn now to seemingly different kind of preliminaries. They will, however, result in another O'Reilly type conclusion. The latter, in turn, will help us to drop certain "variation" conditions used by Goldie (1977) when treating empirical Lorenz processes and also the similar tail conditions of Hall and Wellner (1979) when dealing with mean residual life processes.

For each $s \in (0,1]$, let $F_s^{(n)}$ denote the σ-field generated by the random variables $\{E_n(u) : u \in [s,1]\}$, and let $G_s^{(n)}$ denote the σ-field generated by the random variables $\{B_n(u) : u \in [s,1]\}$. Observe that for each $s \le t$ and $n = 1,2,\ldots$, both $F_t^{(n)} \subset F_s^{(n)}$ and $G_t^{(n)} \subset G_s^{(n)}$ hold true. Introduce

(2.16)
$$\hat{R}_n(s) = \alpha_n(s)/s, \quad s \in (0,1] ,$$
$$R_n(s) = B_n(s)/s, \quad s \in (0,1] .$$

It is well-known and easy to show that both $\{(\hat{R}_n(s), F_s^{(n)}) : s \in (0,1]\}$ and $\{(R_n(s), G_s^{(n)}) : s \in (0,1]\}$ are separable square integrable reverse martingales for each $n = 1,2,\ldots$.

We shall use the following version of Theorem 5.1 of Birnbaum and Marshall (1961).

LEMMA 2.3. Let $\{(Y_t, A_t) : 0 \le t \le 1\}$ be a separable square integrable reverse martingale such that $Y_1 = 0$ a.s. Choose $0 < d < 1$ and let ℓ be a nondecreasing nonnegative function on $[d,1)$ such that

(2.17) $$\int_d^1 \ell^2(s)\,ds < \infty .$$

Then for any $\lambda > 0$,

$$P\{\sup_{d\le t\le 1} |\ell(t)Y_t| \ge \lambda\} \le \lambda^{-2} \int_d^1 \ell^2(s)\,d\mu(s) ,$$

where $\mu(s) = -EY_s^2$.

The following Chibisov (1964)-Pyke and Shorack (1968)-O'Reilly type lemma is a version of a result due to Pyke and Shorack (1968). Such results are usually proved assuming that ℓ is continuous. For the sake of completeness we give our version here including its short proof.

LEMMA 2.4. Let ℓ be a nonnegative function on $(0,1)$ such that there exist two possibly degenerate intervals $(0,a)$ and $(b,1)$, $0 \le a < b \le 1$, so that ℓ is nonincreasing on $(0,a)$ and nondecreasing *on* $(b,1)$, $\sup\{\ell(t) : a \le t \le b\} < \infty$ *and*

$$\int_0^1 \ell^2(t)dt < \infty .$$

Then with B_n *as in* (2.1), (2.6)

$$\Delta_n = \sup_{0<t<1} |\ell(t)\{\alpha_n(t)-B_n(t)\}| \xrightarrow{P} 0.$$

Proof. Choose $c \in (0,a)$ so small that $d = 1-c \in (b,1)$ and notice that

$$\Delta_n \le \sup_{c\le t\le d} |\ell(t)\{\alpha_n(t)-B_n(t)\}|$$

$$+ \sup_{0<t\le c} |\ell(t)\alpha_n(t)| + \sup_{0<t\le c} |\ell(t)B_n(t)|$$

$$+ \sup_{d\le t<1} |\ell(t)\alpha_n(t)| + \sup_{d\le t<1} |\ell(t)B_n(t)|$$

$$= \Delta_n(1) +\ldots+ \Delta_n(5).$$

Since ℓ is bounded on $[c,d]$, $\Delta_n(1) \to 0$ almost surely by (2.1). Since $B(1-t)$, $0 \le t \le 1$, is again a Brownian bridge and since $1-U$ is uniformly distributed on $(0,1)$ if U itself is such a random variable, we have

$$P\{\Delta_n(2) \ge \lambda\} = P\{\sup_{d\le t<1} |\ell(1-t)\alpha_n(t)| \ge \lambda\}$$

and

$$P\{\Delta_n(3) \ge \lambda\} = P\{\sup_{d\le t<1} |\ell(1-t)B(t)| \ge \lambda\}$$

for any $\lambda > 0$. Therefore, observing that

$$\sup_{d\le t<1} |\ell(t)\alpha_n(t)| \le \sup_{d\le t<1} |\ell(t)\hat{R}_n(t)|$$

and

$$\sup_{d\le t<1} |\ell(t)B_n(t)| \le \sup_{d\le t<1} |\ell(t)R_n(t)|$$

with $\hat{R}_n$ and R_n as in (2.16), and on letting $\mu(s) = -ER_n^2(s) = -E(\hat{R}_n(s))^2 = -(1-s)/s$, $\mu'(s) = 1/s^2$, Lemma 2.3 gives

$$P\{\Delta_n(2) +\ldots+ \Delta_n(5) \ge 4\lambda\} \le \frac{2}{\lambda^2}\int_d^1 \frac{\ell^2(s)}{s^2}ds + \frac{2}{\lambda^2}\int_d^1 \frac{\ell^2(1-s)}{s^2}ds$$

$$\leq \frac{2}{(\lambda(1-c))^2}\{\int_{1-c}^{1} \ell^2(s)ds + \int_0^c \ell^2(s)ds\}.$$

The latter upper bound tends to zero if $c \to 0$, and hence the lemma is proved.

Of course, if the nonnegative function ℓ of Lemma 2.4 is also continuous on $(0,1)$ and $q(y) = 1/\ell(y)$ is strictly positive and non-decreasing on $(0,1/2]$ and is symmetric about $1/2$, then Lemma 2.4 is a special case of O'Reilly's theorem, i.e., of Lemma 2.1, since

$$\int_0^{1/2} \frac{1}{t} \exp(-\varepsilon \frac{q^2(t)}{t}) dt = \int_0^{1/2} \frac{1}{q^2(t)} \frac{q^2(t)}{t} \exp(-\varepsilon \frac{q^2(t)}{t}) dt$$

$$\leq C \int_0^{1/2} \frac{1}{q^2(t)} dt < \infty$$

for all $\varepsilon > 0$, where $C = C(\varepsilon) = \sup\{x \exp(-\varepsilon x) : 0 \leq x < \infty\}$.

Since condition (2.5) is necessary for the first term of (2.7) to go to zero in probability, we have the following simple corollary, noted only for convenient reference later on.

LEMMA 2.5. <u>Let</u> $Q = F^{-1}$ <u>be continuous on</u> $[0,1)$. <u>If</u>

$$EX^2 = \int_0^1 Q^2(y)dy < \infty ,$$

<u>then for all</u> $\varepsilon > 0$,

$$\int_{1/2}^{1} \frac{1}{1-t} \exp(- \frac{\varepsilon}{(1-t)Q^2(t)}) dt < \infty .$$

Now we turn to some discussion of condition (2.5). O'Reilly (1974) shows that condition (2.5) is equivalent to εq being an upper class function for all $\varepsilon > 0$ around zero for the standard Weiner process, or the Brownian bridge. This means that for a function $q \in Q^*$ in (2.4) and a Brownian bridge B we have (cf. Proposition 2.1 in O'Reilly (1974)),

(2.18) $\quad pr\{B(t) < \varepsilon q(t),\ t \downarrow 0\} = 1$ for all $\varepsilon > 0$ if and only if (2.5) holds.

If, moreover, $q(t)/t^{\frac{1}{2}}$ is decreasing for small $t > 0$, then (2.18) is known to be equivalent to the more familiar Kolmogorov-Petrovskii-Erdős-Feller criterion (see in Itô-McKean (1965), page 33) that

$$\int_0^{1/2} \frac{q(t)}{t^{3/2}} \exp(-\varepsilon \frac{q^2(t)}{2t}) dt < \infty \quad \text{for all } \varepsilon > 0.$$

For $q \in Q^*$, with Q^* as in (2.4), Shorack (1979) and Shorack and Wellner (1982) write

(2.19) $$q(t) = (t \log\log \tfrac{1}{t})^{\frac{1}{2}} g(t) ,$$

and from the formula

(2.20) $$\int_0^{1/2} \frac{1}{t} \exp(-\varepsilon \frac{q^2(t)}{t})\, dt = \int_0^{1/2} \frac{1}{t(\log \frac{1}{t})^{\varepsilon g^2(t)}}\, dt$$

Shorack (1979) infers the incorrect statement that the finiteness of this integral, i.e., O'Reilly's condition (2.5) is equivalent to

(2.21) $$\lim_{t \to 0} g(t) = \infty .$$

Shorack and Wellner (1982) arrive at this statement by a different (necessarily incorrect) argument. We now give a counterexample to this statement.

EXAMPLE. Set

$$a_n = \exp(-\exp n^4), \quad n = 1,2,\ldots,$$

and define the quantities

$$\ell^2(a_n) = \frac{1}{q^2(a_n)} = \frac{1}{a_n(\log\log \frac{1}{a_n})^{\frac{1}{2}}}, \quad n = 1,2,\ldots .$$

For $t \in [a_{n+1}, a_n]$ we define

$$\frac{1}{q^2(t)} = \ell^2(t) = \begin{cases} \ell^2(a_n) & , \quad b_n \le t \le a_n, \\ \ell^2(b_n) + \dfrac{\ell^2(b_n) - \ell^2(a_{n+1})}{b_n - a_{n+1}}(t - b_n), & a_{n+1} \le t \le b_n, \end{cases}$$

where $a_{n+1} < b_n \le a_n$, and this b_n is chosen so close to a_{n+1} that the area of the triangle T_n should not be bigger than the area of the rectangle R_n:

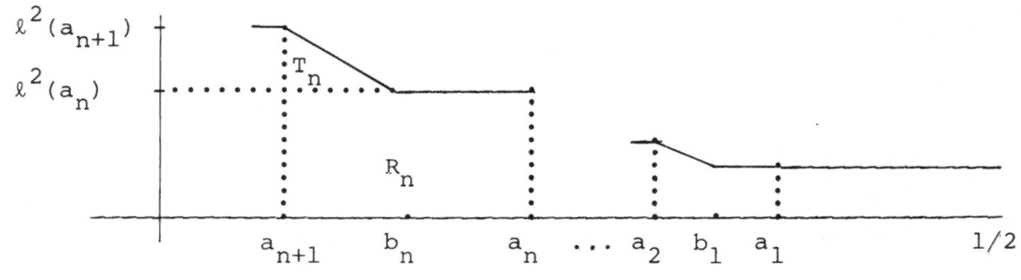

Now $\ell^2(t)$ is defined on $(0, a_1]$, and we set $\ell^2(t) = \ell^2(a_1)$ for $a_1 \le t \le 1/2$. Now $q(t) = 1/\ell(t)$ is continuous and nondecreasing on

(0,1/2], and we have

$$\int_0^{1/2} \frac{1}{q^2(t)}\,dt = \int_0^{1/2} \ell^2(t)\,dt$$

$$\leq \ell^2(a_1)(\tfrac{1}{2} - a_1) + 2\sum_{n=1}^{\infty} \ell^2(a_n)(a_n - a_{n+1})$$

$$= \ell^2(a_1)(\tfrac{1}{2} - a_1) + 2\sum_{n=1}^{\infty} \frac{1}{(\operatorname{loglog} \frac{1}{a_n})^{\frac{1}{2}}} \frac{a_n - a_{n+1}}{a_n}$$

$$\leq \ell^2(a_1)(\tfrac{1}{2} - a_1) + 2\sum_{n=1}^{\infty} \frac{1}{(\operatorname{loglog} \frac{1}{a_n})^{\frac{1}{2}}}$$

$$= \ell^2(a_1)(\tfrac{1}{2} - a_1) + 2\sum_{n=1}^{\infty} \frac{1}{n^2} < \infty$$

and _a fortiori_ condition (2.5) is also satisfied. But for $g(t) = q(t)/(t \operatorname{loglog} t)^{\frac{1}{2}}$ (cf. (2.19)) we have

$$g(a_n) = (\operatorname{loglog} \frac{1}{a_n})^{-\frac{1}{4}} = \frac{1}{n},$$

that is

$$\liminf_{y \to 0} g(y) = 0 .$$

This example means that Shorack's (1979) and Shorack and Wellner's (1982) statement about the equivalency of (2.5) and (2.21) contradicts in fact not only O'Reilly's theorem but also the weaker sufficiency result of Pyke and Shorack (1968), that is, the continuous version of Lemma 2.4 here. Shorack's (1979) two necessity statements for the uniform empirical and O'Reilly's quantile processes are only sufficiency statements. The same statement is formulated about the necessity and sufficiency of (2.21), for the weak approximation version of O'Reilly's result for the uniform empirical process only, in Theorem 1.1 of Shorack and Wellner (1982), where a new proof is given for the sufficiency of (2.21). The error in their proof of the necessity part must be in an incorrect application of an argument of Breiman. It is of interest to point out that the seemingly complety parallel results of Shorack and Wellner (1982) for their uniform empirical process indexed by intervals are true, and the proof of their Proposition 3.1 is correct. Summing up, the equality in (2.19) implies only that

$$\limsup_{t \to 0} g(t) = \infty \quad \text{is necessary for (2.5)},$$

and (2.21) is only a sufficiency condition for (2.5) with a considerable gap in between.

Stute (1982) gives a proof of the sufficiency of (2.5) for (2.*), but he assumes the extra condition that

(2.22) $\frac{q(t)}{t^{1/2}}$ is nonincreasing in a neighbourhood of 0.

That this is indeed an extra assumption follows from the fact that

(2.23) (2.22) and (2.5) imply (2.21).

This means that Stute's extra assumption (2.22) is at least as restrictive as (2.21) of Shorack (1979), and Shorack and Wellner (1982). In fact (2.21) is equivalent to (2.22) and (2.5), because Shorack and Wellner (1982, p. 649) show that (2.21) implies the existence of a $\hat{q} \in Q^*$ such that $\hat{q} \le q$, and the g function of this $\hat{q}$ still goes to ∞ as $t \to 0$, but $\hat{q}(t)/t^{1/2}$ is already nonincreasing on $[0,1/2]$. To show now (2.23), suppose that $t > 0$ is small enough. Then

$$\int_t^{t^{1/2}} \frac{1}{s} \exp\{-\varepsilon \frac{q^2(s)}{s}\} ds \ge \int_t^{t^{1/2}} \frac{1}{s} \exp\{-\varepsilon \frac{q^2(s)}{s}\} ds$$

$$= \exp\{-\varepsilon \frac{q^2(t)}{t}\} \log \frac{1}{t^{1/2}}$$

$$= \frac{1}{2} \exp\{-(\varepsilon \frac{q^2(t)}{t} - \operatorname{loglog} \frac{1}{t})\}.$$

Since the left side upper bound tends to zero as $t \to 0$, we have $m_\varepsilon(t) = \varepsilon(q^2(t)/t) - \operatorname{loglog}(1/t) \to \infty$ as $t \to 0$. But then

$$\frac{q^2(t)}{t} = \frac{1}{\varepsilon}(m_\varepsilon(t) + \operatorname{loglog} \frac{1}{t})$$

$$\ge \frac{1}{\varepsilon} \operatorname{loglog} \frac{1}{t}, \quad 0 < t \le t_\varepsilon,$$

that is, for any small $\varepsilon > 0$ there is a $t_\varepsilon > 0$ such that

$$\frac{q(t)}{(t \operatorname{loglog} \frac{1}{t})^{1/2}} \ge \frac{1}{\varepsilon^{1/2}} \quad \text{for all } 0 < t \le t_\varepsilon .$$

In order to better understand what an O'Reilly weight function is like on the tails, we write

(2.24) $q(t) = t^{1/2}h(t), \quad 0 < t \le 1/2.$

Theorem 2 of Chibisov (1964), or Example 1 of Jaeschke (1979), implies that

$$\sup_{0<t<1} \frac{|\alpha_n(t)|}{(t(1-t))^{1/2}} \xrightarrow{P} \infty,$$

and this of course suggests that $\lim_{t\to 0} h(t) = \infty$. Indeed, the latter is contained in the proof of Proposition 2.1 in O'Reilly (1974, p. 644). For easy reference we formulate this result and give also O'Reilly's simple proof.

LEMMA 2.6. *If* $q(t) = t^{\frac{1}{2}}h(t)$ *is an O'Reilly weight function, that is,* $q \in Q^*$ *and* (2.5) *is satisfied, then*

$$\lim_{t\to 0} h(t) = \infty .$$

Proof. Fixing $0 < \varepsilon < 1$ we have for any $t \in (0,1/2]$

$$\int_{\varepsilon t}^{t} \frac{1}{s} \exp\{-\varepsilon \frac{q^2(s)}{s}\}ds \geq \exp\{-\frac{q^2(t)}{t}\} \int_{\varepsilon t}^{t} \frac{1}{s} ds$$

$$= (\log \frac{1}{\varepsilon}) \exp\{-\frac{q^2(t)}{t}\} > 0$$

using only that q is nondecreasing on $[0,1/2]$. By (2.5) the left side converges to zero as $t \to 0$ implying that $q^2(t)/t \to \infty$ as $t \to 0$.

It is easy to see, but we nevertheless show now that the rate at which h of (2.24) converges to ∞ as $t \to 0$ can never be specified, even if the reciprocal of q is square-integrable. The h function in (2.24) depends on q and may converge to ∞ as slowly and oscillating in general as we wish, that is nothing more than Lemma 2.6 can be said about it.

IMPROVED EXAMPLE. *For any given continuous function* $h^*(\cdot)$ *on* $[0,1/2]$ *such that* $\lim_{t\to 0} h^*(t) = \infty$ *there exists an O'Reilly weight function* q *such that* $\int_0^{1/2} (1/q^2(t))dt < \infty$ *and*

$$\liminf_{t\to 0} \frac{q(t)}{t^{\frac{1}{2}}h^*(t)} = 0 .$$

For constructing the required q, set first

$$\hat{h}(t) = \inf\{h^*(s) : 0 < s \leq t\} \leq h^*(t), \quad 0 < t \leq 1/2.$$

This $\hat{h}$ is nonincreasing and still converges to ∞ as $t \to 0$. Next choose a function $\tilde{h}(t) \leq \hat{h}(t)$ such that $\tilde{h}$ is nonincreasing, converges to ∞ as $t \to 0$, but $t\tilde{h}(t)$ is nondecreasing. That such an $\tilde{h}$ exists on $(0,1/2]$ follows from the fact that for any continuous nondecreasing function $\hat{f}(t) = 1/\hat{h}(t)$, converging to zero as $t \to \infty$, one can find a <u>concave</u> continuous nondecreasing function $\tilde{f}(t) = 1/\tilde{h}(t)$ such that $\hat{f}(t) \leq \tilde{f}(t)$ and $\tilde{f}(t) \to 0$ as $t \to 0$. Since $\tilde{f}$ is concave, $\tilde{f}(t)/t = 1/(t\tilde{h}(t))$ is nonincreasing, and we have the required $\tilde{h}$. Define now the a_n sequence as

$$a_n = \inf\{t : \tilde{h}(t) \geq n^2\},$$

and let n_o be the smallest n for which $a_{n_o} < 1/2$. If $n \geq n_o$ then $a_{n+1} < a_n$, and for such n we now put

$$\ell^2(a_n) = \frac{1}{q^2(a_n)} = \frac{1}{a_n\tilde{h}(a_n)}, \quad n \geq n_o .$$

For $t \in [a_{n+1}, a_n]$, $n \geq n_o$, we define $\ell^2(t) = 1/q^2(t)$ exactly as in the Example above, in terms of the new a_n sequence, and, finally, we let $\ell^2(t) = 1/q^2(t) = \ell^2(a_{n_o})$ for $a_{n_o} \leq t \leq 1/2$. Clearly $q(\cdot) = 1/\ell(\cdot) \in Q^*$, and

$$\int_0^{1/2} \frac{1}{q^2(t)}\, dt \leq \ell^2(a_{n_o})(\tfrac{1}{2} - a_{n_o}) + 2 \sum_{n=n_o}^{\infty} \frac{a_n - a_{n+1}}{a_n\tilde{h}(a_n)}$$

$$\leq \ell^2(a_{n_o})(\tfrac{1}{2} - a_{n_o}) + 2 \sum_{n=n_o}^{\infty} \frac{1}{n^2} < \infty .$$

But

$$\frac{q(a_n)}{a_n^{1/2}h^*(a_n)} \leq \frac{q(a_n)}{a_n^{1/2}\tilde{h}(a_n)} = \frac{1}{n},$$

and this implies (2.23) in view of the fact that $a_n \to 0$ as $n \to \infty$.

Having now completed our discussion of O'Reilly's theorems in terms of square-integrable $(1/q)$ functions, we still need to list a few more preliminaries for later use. We shall need some inequalities of Wellner (1978) for the ratios $t/E_n(t)$, $t/U_n(t)$ and their reciprocals. In fact the special inequalities contained in Wellner's (1978), Remark 1, will suffice for us, and we formulate these as follows. The equality in the first line is due to Daniels (1945).

LEMMA 2.7. <u>For all</u> $\lambda \geq 1$

$$P\{\sup_{1/n \leq t \leq 1} \frac{t}{U_n(t)} \geq \lambda\} \leq P\{\sup_{0 \leq t \leq 1} \frac{E_n(t)}{t} \geq \lambda\} = \frac{1}{\lambda}$$

<u>and</u>

$$P\{\sup_{1/n \leq t \leq 1} \frac{U_n(t)}{t} \geq \lambda\} \leq P\{\sup_{U_{1:n} \leq t \leq 1} \frac{t}{E_n(t)} \geq \lambda\} \leq e\lambda e^{-\lambda} .$$

In a certain sense the following result of Csáki (1977) may be viewed as a "strong version" of the Daniels equality above. This result is Csáki's (1977) Theorem 3.2.

LEMMA 2.8. <u>If</u> $\delta_n(d) = dn^{-1}\log\log n$ <u>with some constant</u> $d > 0$,

<u>then</u>

$$\limsup_{n\to\infty} \sup_{\delta_n(d)\le x\le 1-\delta_n(d)} \frac{|\alpha_n(x)|}{(x(1-x)\log\log n)^{\frac12}} = \ell_d := \max(2, d^{\frac12}(\beta_d - 1))$$

<u>almost surely, where</u> $\beta_d > 1$ <u>is the solution of the equation</u> $\beta(\log\beta - 1) = d^{-1}(1-d)$. <u>Specifically, if</u> $d = 1$ <u>then</u> $\ell_1 = 2$.

Applying this result of Csáki with $d = 0{,}236\ldots$, for which $d^{\frac12}(\beta_d - 1) = 2$, M. Csörgő and Révész (1978; see also 1981, Theorem 4.5.5) derived that

$$\limsup_{n\to\infty} \sup_{\delta_n(25)\le x\le 1-\delta_n(25)} \frac{|u_n(x)|}{(x(1-x)\log\log n)^{\frac12}} \le 4 \quad \text{a.s.} \tag{2.25}$$

The latter result in turn easily implies (see top of p. 151 of M. Csörgő and Révész (1981)) the following.

LEMMA 2.9.

$$\limsup_{n\to\infty} \sup_{\delta_n(25)\le t\le 1} \frac{t}{U_n(t)} \le 6 \quad \underline{\text{a.s.}}$$

<u>and</u>

$$\limsup_{n\to\infty} \sup_{1-\delta_n(25)\le t\le 1} \frac{1-t}{1-U_n(t)} \le 6 \quad \underline{\text{a.s.}}$$

The intervals $[\delta_n(25), 1-\delta_n(25)]$ almost surely do not cover the largest and smallest uniform order statistics $U_{n:n}$ and $U_{1:n}$, and in our strong approximation proofs we shall frequently need to know the order of these variables. This is the content of the following well-known estimates.

LEMMA 2.10.

$$\liminf_{n\to\infty} (\log n)^2 nU_{1:n} = \infty \quad \underline{\text{a.s.}},$$

$$\limsup_{n\to\infty} (\log n)^{-1} nU_{1:n} < \infty \quad \underline{\text{a.s.}},$$

<u>and</u>

$$\liminf_{n\to\infty} (\log n)^2 n(1-U_{n:n}) = \infty \quad \underline{\text{a.s.}},$$

$$\limsup_{n\to\infty} (\log n)^{-1} n(1-U_{n:n}) < \infty \quad \underline{\text{a.s.}}$$

Here the limsup statements follow from the more general upper-upper class result of Robbins and Siegmund (1972), while the liminf statements follow from the more general lower-lower class result of Geffroy (1958/59).

3. AUXILIARY PROCESSES: INTEGRALS OF EMPIRICAL PROCESS.

LEMMA 3.1. *If* $\mu = \int_0^1 Q(y)dy < \infty$, *then*

$$\Delta_n^{(1)} = \sup_{0<u<1} \left|\int_0^u (1-E_n(v))dQ(v) - \int_0^u (1-v)dQ(v)\right| \xrightarrow{a.s.} 0$$

Proof. Let $\varepsilon > 0$ be arbitrarily small and choose $\beta \in (0,1)$ so small that

$$I^{(3)}(\beta) = \int_{1-\beta}^1 (1-v)dQ(v) < \varepsilon/2.$$

Then

$$\Delta_n^{(1)} \le I_n^{(1)}(\beta) + I_n^{(2)}(\beta) + I^{(3)}(\beta),$$

where, with $\chi(A)$ denoting the indicator of an event A ,

$$I_n^{(1)}(\beta) = n^{-1}\sum_{i=1}^n \int_{1-\beta}^1 \chi(\{U_i > v\})dQ(v) \xrightarrow{a.s.} I(\beta)$$

by the strong law of large numbers, and

$$I_n^{(2)}(\beta) = \sup_{0<u\le 1-\beta} \left|\int_0^u (1-E_n(v))dQ(v) - \int_0^u (1-v)dQ(v)\right|$$

$$= \sup_{0<u\le 1-\beta} \left|\int_0^u (v-E_n(v))dQ(v)\right|$$

$$\le Q(1-\beta) \sup_{0\le u\le 1} |u-E_n(u)| \xrightarrow{a.s.} 0$$

by Glivenko-Cantelli. Thus $\limsup_{n\to\infty} \Delta_n^{(1)} \le \varepsilon$ for all $\varepsilon > 0$ and the lemma is proved.

When strongly approximating the process

$$(3.1) \qquad \beta_n(u) = n^{\frac{1}{2}}\{\int_0^u (1-E_n(v))dQ(v) - H_F^{-1}(u)\}$$

in Lemma 3.1, with the definition of H_F^{-1} as in (1.5), conditions of the form

$$(3.2) \qquad J(r) = \int_0^1 (1-u)^{1/r}dQ(u) = \int_0^\infty (1-F(x))^{1/r}dx < \infty ,$$

for $r \ge 2$, will play an important role. This is slightly stronger than the existence of the rth moment. Indeed, on extending the discussion in the Appendix of Hoeffding (1973), we see that $J(r) < \infty$ implies $EX^r < \infty$. This is not necessarily true conversely, but $E\{X^r(\log(1+X))^{1+\delta}\} < \infty$, with any $\delta > 0$, already implies $J(r) < \infty$. The integral

$$\int_0^1 B(v)\,dQ(v) = \int_0^\infty B(F(x))\,dx$$

with a Brownian bridge B will emerge in a natural way in all our subsequent processes, and it will be very convenient sometimes to regard these integrals as proper Lebesgue integrals rather than as improper Riemann integrals. In order to further motivate the $J(r)$ conditions we note that

$$(3.3) \qquad \int_0^\infty |B(F(x))|\,dx < \infty \quad \text{a.s. if and only if} \quad J(2) < \infty ,$$

since

$$E\int_0^\infty |B(F(x))|\,dx = \int_0^\infty E|B(F(x))|\,dx = \left(\tfrac{2}{\pi}\right)^{\frac{1}{2}}\int_0^\infty (F(x)(1-F(x)))^{\frac{1}{2}}dx.$$

On the other hand, the integral in question may exist as an improper Riemann integral under weaker conditions. We show that

$$(3.4) \qquad EX^2 < \infty \quad \text{implies} \quad P\{\sup_{0\le u\le 1} |\int_0^u B(y)\,dQ(y)| < \infty\} = 1.$$

Indeed, integrating by parts,

$$\int_0^u B(y)\,dQ(y) = B(u)Q(u) - \int_0^u Q(y)\,dB(y), \quad 0 \le u \le 1,$$

provided that the latter stochastic integral (defined through the distributional equality $B(y) = W(y)-yW(1)$ with a Wiener process W) exists. Since Q is square integrable on $[0,1]$, it indeed exists. But then this stochastic integral is almost surely continuous, as a function of u on $[0,1]$, by Theorem 3 of §2 in Chapter 2 of Skorohod (1965). On the other hand,

$$(3.5) \qquad EX^2 < \infty \quad \text{implies} \quad P\{\sup_{0\le u<1} |B(u)Q(u)| < \infty\} = 1$$

by the proof of Lemma 2.4, i.e., by a simple application of the Birnbaum-Marshall inequality in Lemma 2.3, or directly by Lemma 2.5 and (2.18).

Now for β_n in (3.1) and B_n in (2.6) we have the following result.

LEMMA 3.2. *If* $EX^2 < \infty$ *then*

$$\Delta_n^{(2)} = \sup_{0\le u\le 1} |\beta_n(u) - \int_0^u B_n(y)\,dQ(y)| \xrightarrow{P} 0 .$$

Proof. Using the definitions of β_n and H_F^{-1} we obtain by simple manipulation that

$$\Delta_n^{(2)} = \sup_{0\le u\le 1} \left|\int_0^u (\alpha_n(y)-B_n(y))\,dQ(y)\right|$$

$$\le 2\int_0^{1-\varepsilon} |\alpha_n(y)-B_n(y)|\,dQ(y)+I_n^{(3)}(\varepsilon)+I_n^{(4)}(\varepsilon)$$

for any $\varepsilon \in (0,1)$, where

$$I_n^{(3)}(\varepsilon) = \sup_{1-\varepsilon\le u\le 1} \left|\int_{1-\varepsilon}^u B_n(y)\,dQ(y)\right| \xrightarrow{P} 0$$

as $\varepsilon \to 0$ by (3.4), on noting that $I_n^{(3)}(\varepsilon)$ has the same distribution for each n, and

$$I_n^{(4)}(\varepsilon) = \sup_{1-\varepsilon\le u\le 1} \left|\int_{1-\varepsilon}^u \alpha_n(y)\,dQ(y)\right|.$$

Introduce now the processes

$$\hat{I}_n(u) = \int_u^1 \hat{R}_n(y)\,dQ(y)+\hat{R}_n(u)Q(u), \quad u \in (0,1],$$

where $\hat{R}_n$ is defined in (2.16) and we also recall the definition of $F_u^{(n)}$ from there. It is routine to establish that $\{(\hat{I}_n(u),F_u^{(n)}): u \in (0,1]\}$ is a separable square integrable reverse martingale for each n. Notice also that by the c_r-inequality (Loève (1960), p.155)

$$E(\hat{I}_n(u))^2 \le 2\int_u^1\int_u^1 (\min(s,t)-st)\,dQ(s)\,dQ(t) + 2u(1-u)Q^2(u)$$

(3.6)

$$\le 4\int_u^1 Q^2(y)\,dy.$$

Now, with any $u \in [1-\varepsilon,1]$, for the processes in $I_n^{(4)}(\varepsilon)$ we have

$$\int_{1-\varepsilon}^u \alpha_n(y)\,dQ(y) = \int_{1-\varepsilon}^u \hat{R}_n(y)\,y\,dQ(y)$$

$$= \int_{1-\varepsilon}^u y\,\frac{d}{dy}\left\{-\int_y^1 \hat{R}_n(s)\,dQ(s)\right\}dy$$

$$= \left[-y\int_y^1 \hat{R}_n(s)\,dQ(s)\right]_{1-\varepsilon}^u + \int_{1-\varepsilon}^u\int_y^1 \hat{R}_n(s)\,dQ(s)\,dy$$

$$= (1-\varepsilon)\hat{I}_n(1-\varepsilon)-(1-\varepsilon)\hat{R}_n(1-\varepsilon)Q(1-\varepsilon)$$

$$- u\hat{I}_n(u) + u\hat{R}_n(u)Q(u)$$

$$+ \int_{1-\varepsilon}^u \hat{I}_n(y)\,dy - \int_{1-\varepsilon}^u \hat{R}_n(y)Q(y)\,dy\,.$$

Hence

$$I_n^{(4)}(\varepsilon) \le 3 \sup_{1-\varepsilon\le u\le 1} |\hat{I}_n(u)| + (2+(1-\varepsilon)^{-1}) \sup_{1-\varepsilon\le u\le 1} |B_n(u)Q(u)|.$$

Since by Lemma 2

$$P\{\sup_{1-\varepsilon\le u\le 1} |\hat{I}_n(u)| > \lambda\} \le -\lambda^{-2} \int_{1-\varepsilon}^{1} d(E\hat{I}_n(u))^2$$
$$= \lambda^{-2}E(\hat{I}_n(1-\varepsilon))^2$$
$$\le \frac{4}{\lambda^2}\int_{1-\varepsilon}^{1} Q^2(y)\,dy$$

for any $\lambda > 0$, we obtain

$$\lim_{\varepsilon\to 0} \limsup_{n\to\infty} P\{I_n^{(4)}(\varepsilon) > \lambda\} = 0,$$

on applying also (3.5) for the second term of the bound in (3.7). This completes the proof.

Next we prove a strong version of Lemma 3.2.

LEMMA 3.3. *If* $J(r) < \infty$ *for* $r > 2$, *then*

$$\Delta_n^{(2)} = \sup_{0<u<1} |\beta_n(u) - \int_0^u B_n(v)\,dQ(v)| \overset{a.s.}{=} O(n^{-\lambda})$$

for any $\lambda \in (0, \frac{1}{2} - \frac{1}{r})$.

Proof. Let $\varepsilon_n = n^{-\tau}$ with $0 < \tau < 1$. The condition that $J(r) < \infty$ implies that

$$Q(1-\varepsilon_n) \le \varepsilon_n^{-1/r} = n^{\tau/r}$$

for all large enough n. Hence using (2.1), by the first steps of the proof of Lemma 3.2 we obtain

$$\Delta_n^{(2)} \le \int_0^{1-\varepsilon_n} |\alpha_n(v)-B_n(v)|\,dQ(v) + I_n^{(3)}(\varepsilon_n) + I_n^{(4)}(\varepsilon_n)$$
$$\overset{a.s.}{=} O((\log n)^2 n^{\alpha/r-1/2}) + I_n^{(3)}(\varepsilon_n) + I_n^{(4)}(\varepsilon_n).$$

Let $\delta > 0$. Using (2.8) in the third step of the proof of Lemma 3.2, we obtain

$$I_n^{(3)}(\varepsilon_n) \le \sup_{1-\varepsilon_n\le y<1} \frac{|\alpha_n(y)|}{(1-y)^{\frac{1}{2}-\delta}} \int_{1-\varepsilon_n}^{1} (1-v)^{\frac{1}{2}-\delta}\,dQ(v)$$

$$\leq \varepsilon_n^{\delta/2} \sup_{1-\varepsilon_n \leq y < 1} \frac{|\alpha_n(y)|}{(1-y)^{\frac{1}{2}-\delta/2}} \int_{1-\varepsilon_n}^{1} (1-v)^{\frac{1}{2}-\delta} dQ(v)$$

$$\stackrel{a.s.}{=} O(\varepsilon_n^{\delta/2} (\log\log n)^{\frac{1}{2}}) \int_{1-\varepsilon_n}^{1} (1-v)^{\frac{1}{r} + (\frac{1}{2} - \frac{1}{r} - \delta)} dQ(v)$$

$$\leq O(1) \varepsilon_n^{1/2-1/r-\delta} J(r)$$

$$= O(n^{-\tau(1/2-1/r-\delta)}).$$

Now using (2.9) instead of (2.8), we obtain in exactly the same way that

$$I_n^{(4)}(\varepsilon_n) = \int_{1-n^{-\tau}}^{1} |B_n(v)| dQ(v) \stackrel{a.s.}{=} O(n^{-\tau(1/2-1/r-\delta)}). \tag{3.7}$$

This also completes the proof of Lemma 3.3, since $\delta > 0$ is arbitrarily small and $\tau < 1$ can be as close to 1 as we wish.

4. MEAN RESIDUAL LIFE PROCESSES.

We summarise now a convergence theory for the mean residual life process z_n of (1.15) as a consequence of the preceding section. Clearly,

$$z_n(x) = (1-F_n(x))^{-1} \int_{F(x)}^{1} \alpha_n(y)dQ(y)$$
$$+ (1-F_n(x))^{-1}M_F(Q(F(x)))n^{\frac{1}{2}}(F_n(x)-F(x)),$$

and hence its approximating Gaussian process will be

$$Z_n(x) = (1-F(x))^{-1} \int_{F(x)}^{1} B_n(y)dQ(y)$$
$$-(1-F(x))^{-1}M_F(Q(F(x)))B_n(F(x)).$$

Setting $T_F = \inf\{t:F(t)=1\}$, we have the following result.

THEOREM 4.1. (i) *If* $\mu < \infty$ *and* $T < T_F$, *then*

$$\sup_{0\le t\le T} |M_n(t)-M_F(t)| \xrightarrow{a.s.} 0 .$$

(ii) *If* $EX^2 < \infty$ *and* $T < T_F$, *then*

$$\sup_{0\le t\le T} |z_n(t)-Z_n(t)| \xrightarrow{P} 0.$$

(iii) *If* $J(r) < \infty$ *for some* $r > 2$ *and* $T < T_F$ *then*

$$\sup_{0\le t\le T} |z_n(t) - Z_n(t)| \stackrel{a.s.}{=} O(n^{-\lambda})$$

for any $\lambda \in (0, \frac{1}{2} - \frac{1}{r})$.

(iv) *If* $EX^2 < \infty$ *then*

$$\Delta_n^{(3)} = \sup_{0\le x<\infty} |(1-F_n(x))z_n(x)-(1-F(x))Z_n(x)| \xrightarrow{P} 0.$$

(v) *If* $J(r) < \infty$ *for some* $r > 2$, *then*

$$\Delta_n^{(3)} = \sup_{0\le x<\infty} |(1-F_n(x))z_n(x)-(1-F(x))Z_n(x)| \stackrel{a.s.}{=} O(n^{-\lambda})$$

for any $\lambda \in (0, \frac{1}{2} - \frac{1}{r})$.

Proof. Part (i) follows from Lemma 3.1 trivially. Part (ii) follows directly from Lemma 3.2 and (2.1). Part (iii) follows again directly from Lemma 3.3 and (2.1). To prove (iv) and (v), we note that

$$\Delta_n^{(3)} \leq \sup_{0\leq x<\infty} \left|\int_{F(x)}^{1} (\alpha_n(y)-B_n(y))\,dQ(y)\right| \tag{4.1}$$
$$+ \sup_{0\leq x<\infty} \left|M_F(Q(F(x)))\{\alpha_n(F(x))-B_n(F(x))\}\right|.$$

Under the existence of the variance the first term goes to zero in probability by Lemma 3.2, while the second term is of the form

$$\sup_{0<t<1} |\ell(t)\{\alpha_n(t)-B_n(t)\}|$$

of Lemma 2.4. Hence this will also converge to zero in probability if we show that the function

$$\begin{aligned}\ell(t) &= M_F(Q(t))\\ &= (1-t)^{-1}\{(1-t)Q(t) + \int_t^1 Q(y)\,dy\\ &= Q(t) + (1-t)^{-1} \int_t^1 Q(y)\,dy\end{aligned}$$

is nondecreasing and square integrable on $(0,1)$. The first term, $Q(t)$, satisfies these conditions, and therefore it is enough to show them for the second term

$$\ell^*(t) = (1-t)^{-1} \int_t^1 Q(y)\,dy.$$

We have

$$\begin{aligned}\frac{d}{dt}\,\ell^*(t) &= -\frac{Q(t)}{1-t} + \frac{\int_t^1 Q(y)\,dy}{(1-t)^2}\\ &= (1-t)^{-2}\{\int_t^1 Q(y)\,dy-(1-t)Q(t)\}\end{aligned}$$

almost everywhere in $(0,1)$, and the latter is nonnegative by the monotonicity of Q. On the other hand,

$$\begin{aligned}\int_0^1 (\ell^*(t))^2 dt &= \int_0^1 (\ell^*(1-t))^2 dt\\ &= \int_0^1 (\tfrac{1}{t} \int_0^t Q(1-y)\,dy)^2 dt\\ &\leq 2 \int_0^1 Q^2(1-y)\,dy\\ &= 2 \int_0^1 Q^2(y)\,dy < \infty\end{aligned}$$

by Hardy's inequality (Rudin (1966), p.72). Thus part (iv) is proved.

We note that the square integrability of ℓ^* clearly implies the square integrability of Q, that is, the existence of the second moment and the square integrability of ℓ^* are in fact equivalent.

Towards completing now the proof of part (v), we note that $J(r) < \infty$, $r > 2$, implies that the first term on the right side of (4.1) is a.s. $O(n^{-\lambda})$ by an application of Lemma 3.3.

It is therefore enough to show that the second term above is also a.s. $O(n^{-\lambda})$. As we have already noted, $J(r) < \infty$ implies $EX^r < \infty$. The latter, in turn, according to Proposition 1(c) of Hall and Wellner (1981), implies that

$$M_F(x) < EX^r(1-F(x))^{-1/r} - x$$

for all $x > 0$. Hence $J(r) < \infty$ implies

$$C = \sup_{0 \leq x < \infty} (1-F(x))^{1/r} M_F(x) < \infty.$$

The second term above is therefore not greater than

$$C\{\sup_{0<y\leq 1-\varepsilon_n} \frac{|\alpha_n(y)-B_n(y)|}{(1-y)^{1/r}} + \sup_{1-\varepsilon_n\leq y<1} \frac{|\alpha_n(y)|}{(1-y)^{1/r}} + \sup_{1-\varepsilon_n\leq y<1} \frac{|B_n(y)|}{(1-y)^{1/r}}\},$$

with $\varepsilon_n = n^{-\alpha}$, $0 < \alpha < 1$, of the proof of Lemma 3.3. The first term in the last bracket is again a.s.

$$O((\log n)^2\, n^{\alpha/r-1/2}).$$

The second term is, with $\delta > 0$,

$$\sup_{1-\varepsilon_n\leq y<1} \frac{|\alpha_n(y)|}{(1-y)^{\frac{1}{2}-\delta+(\frac{1}{r}-\frac{1}{2}+\delta)}} \leq \varepsilon_n^{\frac{1}{2}-\frac{1}{r}-\delta} \sup_{1-\varepsilon_n\leq y<1} \frac{|\alpha_n(y)|}{(1-y)^{\frac{1}{2}-\delta}}$$

$$\overset{\text{a.s.}}{=} O(n^{-\alpha(\frac{1}{2}-\frac{1}{r}-\delta)})$$

by (2.8), and we obtain the same $O(\cdot)$ rate for the third term by applying (2.9). Hence part (v) is also proved.

We should point out that the consistency result in (i) was proved by Yang (1978). Part (ii) was first proved also by Yang (1978) under the additional assumption that the density function f of F exists and is positive on the support of F. Hall and Wellner (1979) noted that a careful inspection of her proof reveals that this assumption is superfluous. An analogue of (iii) was derived directly from (2.1) by Burke, S. Csörgő and Horváth (1981) under the assumption of finite life, i.e., $T_F < \infty$. The rate of the approximation is $O((\log n)^2/n^{\frac{1}{2}})$ in this case. Part (iv) is a slightly stronger weak convergence result than the corresponding one of Hall and Wellner (1979) since they use a weaker approximation result of Shorack instead of the weak approximation version of the Pyke and Shorack (1968) or the O'Reilly theorem, i.e., the first half of Lemma 2.1, and they do not have a result to handle directly the second term in the bound of (4.1) as we did.

Part (v) implies a functional and an ordinary law of the iterated logarithm for

$$(4.2) \qquad v_n(x) = (1-F_n(x))z_n(x), \quad 0 \le x < \infty,$$

and for its absolute supremum under the slightly stronger, but perhaps simpler condition $J(r) < \infty$, $r > 2$, than those of Hall and Wellner (1979). On the other hand, a strong approximation result is a richer statement than a log log law. An important observation of Hall and Wellner (1979) identifies the weak limit of v_n in (4.2) as a constant multiple of a scaled Wiener process. Their observation implies that for our Gaussian processes $V_n(x) = (1-F(x))Z_n(x)$, $0 \le x < \infty$, we have

$$EV_n(x)V_m(t) = \frac{n \wedge m}{(nm)^{\frac{1}{2}}} \sigma^2(0)(\tilde{R}(x) \wedge \tilde{R}(t)),$$

where

$$\tilde{R}(x) = (1-F(x))\sigma^2(x)/\sigma^2(0)$$

with

$$\sigma^2(x) = \mathrm{Var}(X-x \mid X > x),$$

and $\tilde{G}(x) = 1-\tilde{R}(x)$ is a distribution function on $[0,\infty)$. Since $EV_n(x) \equiv 0$, this means that the equality

$$\{V_n(x),\ 0 \le x < \infty,\ n=1,2,\ldots\} \stackrel{\mathcal{D}}{=} \{\sigma(0)n^{-\frac{1}{2}}W(\tilde{R}(x),n),\ 0 \le x < \infty,\ n=1,2,\ldots\}$$

in distribution holds, where $W(\cdot,\cdot)$ is a standard two-parameter Wiener process on the non-negative quadrant of the plane. Suppose now that the density function f of F exists. Then $\tilde{G}$ also has a density $\tilde{g}(x) = f(x)M^2(x)/\sigma^2(0)$, for which we assume that $C = \sup\{g(x): 0 < x < T_F\} < \infty$. Let $h_n \in (0,1]$ be a non-increasing sequence of constants and set

$$\gamma_n = (2Ch_n\{\log\log n + \log \frac{1}{Ch_n}\})^{-\frac{1}{2}}.$$

If T_n is such a sequence of constants that $T_n \ge C^{-1}(Kh_n)$, then

$$(4.3) \qquad \limsup_{n\to\infty} \gamma_n \sup_{T_n \le t < \infty} \sup_{0 < u \le h_n} |v_n(t+u) - v_n(t)| \le \sigma(0),$$

provided that h_n is such that $\gamma_n n^{-\lambda} \to 0$ with λ as in (v). This strong law for the fluctuation of the increments of the by $(1-F_n)$ multiplied mean residual life process follows from (v) and the corresponding result for the Wiener sheet, as given by Chan's Theorem S.1.14.2 in M. Csörgő and Révész (1981), by a simple elaboration on the scaled sheet in (4.2).

In order to further motivate the strength of the approximation method and to accompany the log log of Hall and Wellner (1979) for the limsup referred to above, we note that Chung's (1948) other law of the iterated logarithm (cf. also Jain and Pruitt (1975)) implies that

$$\liminf_{n\to\infty} \left(\frac{\log\log n}{n}\right)^{1/2} \sup_{0\le x<\infty} |W(\tilde{R}(x),n)| = \frac{\pi}{\sqrt{8}} .$$

Hence, if $J(r) < \infty$ for some $r > 2$, then Part (v) of Theorem 4.1 implies that

$$\liminf_{n\to\infty} (\log\log n)^{1/2} \sup_{0\le x<\infty} |v_n(x)| = \frac{\pi}{\sqrt{8}}\,\sigma(0) .$$

5. AUXILIARY PROCESSES: EMPIRICAL INCREMENTS OF BROWNIAN BRIDGE INTEGRALS.

An implication of the following lemma will be useful.

LEMMA 5.1. *If* $EX^2 < \infty$ *and* $Q = F^{-1}$ *is continuous on* $[0,1)$ *then the stochastic process*

$$\int_0^u B(y)\,dQ(y)$$

is almost surely continuous on $[0,1]$.

Proof. As in the proof of (3.4), our process is

$$B(u)Q(u) - \int_0^u Q(y)\,dB(y), \quad 0 \le u \le 1, \tag{5.1}$$

where, as we saw there, the second term is almost surely continuous on $[0,1]$, while the first term is almost surely continuous on $[0,1)$. On defining, however, this first term of (5.1) by zero at $u=1$, the left continuity at $u=1$ also follows by the Birnbaum-Marshall inequality in Lemma 2.3 as used in the proof of Lemma 2.4, or directly by Lemma 2.5 and (2.18).

For U_n and B_n as in (2.2) and (2.6), respectively, this lemma readily implies the following.

LEMMA 5.2. *If* $EX^2 < \infty$ *and* $Q = F^{-1}$ *is continuous on* $[0,1)$, *then*

$$\Delta_n^{(4)} = \sup_{0\le y\le 1} \left| \int_0^{U_n(y)} B_n(u)\,dQ(u) - \int_0^y B_n(u)\,dQ(u) \right| \xrightarrow{P} 0.$$

In order to produce now a strong version of Lemma 5.2, we have to be able to measure the increments of Q. It is then natural to assume the existence of such a density $f = F'$ which is <u>everywhere</u> positive on the support of F. Of course, even then the derivative $Q' = 1/f(Q)$ - the quantile-density function - cannot grow arbitrarily around 0 and 1, and disregarding the logarithmic factors, we shall require the following condition

$$J\left(\frac{1}{\alpha}, \frac{1}{\beta}\right) = \sup_{0<u<1} \frac{u^{\alpha}(1-u)^{\beta}}{f(Q(u))} < \infty \tag{5.2}$$

for some $\alpha, \beta \in [0,3/2)$. We shall see later on that an even more restrictive growth condition will in fact be "almost" <u>necessary</u> for the mere weak convergence of the total time on test process. The number $3/2$ is not curious here because a simple manipulation shows that

(5.3) $\quad J(1/\alpha,1/\beta) < \infty$ implies $J(1/\gamma) < \infty \begin{cases} \text{for any } \gamma > \beta-1 & \text{if } \beta \geq 1, \\ \text{for any } \gamma > 0 & \text{if } 0 \leq \beta < 1, \end{cases}$

with any $\alpha,\beta \geq 0$, and $\beta < 3/2$ is the same as $1/(\beta-1) > 2$. Since $F(0) = 0$, with $C_1, C_2, \ldots$ denoting appropriate constants we have

$$Q(u) \leq C_1, \quad 0 \leq u \leq 1/2, \tag{5.4}$$

and condition (5.2) implies

$$Q(u) \leq C_2 u^{1-\alpha}, \quad 0 \leq u \leq 1/2, \quad \text{if} \quad \alpha < 1 \tag{5.5}$$

and

$$Q(u) \leq \begin{cases} C_3 & , \quad 1/2 \leq u \leq 1, \quad \text{if} \quad \beta < 1 \\ C_4 \log \frac{1}{1-u} & , \quad 1/2 \leq u \leq 1, \quad \text{if} \quad \beta = 1 \\ C_5 (1-u)^{1-\beta} & , \quad 1/2 \leq u \leq 1, \quad \text{if} \quad \beta > 1 \end{cases} \tag{5.6}$$

by a simple integration argument

LEMMA 5.3. *If the density function* $f(x) = F'(x)$ *is positive on the open support of* F *and* $J(1/\alpha,1/\beta) < \infty$ *for some* $\alpha,\beta \in [0,3/2)$, *then*

$$\Delta_n^{(4)} = \sup_{0<y<1} \left| \int_0^{U_n(y)} B_n(u)\,dQ(y) - \int_0^y B_n(u)\,dQ(u) \right| \overset{a.s.}{=} O(n^{-\rho})$$

for any $\rho < \min(\nu(\alpha), \frac{3-2\beta}{10-4\beta})$, *where*

$$\nu(\alpha) = \begin{cases} \frac{1}{6} & , \quad \text{if} \quad \alpha \geq 1, \\ \frac{3-2\alpha}{10-4\alpha} & , \quad \text{if} \quad \alpha < 1 . \end{cases}$$

Proof. Set $\varepsilon_{1n} = n^{-\tau_1}$ and $\varepsilon_{2n} = n^{-\tau_2}$ where

$$0 < \tau_1 = \tau_1(\alpha) < \begin{cases} \frac{1}{3} & , \quad \text{if} \quad \alpha \geq 1, \\ \frac{1}{5-2\alpha} & , \quad \text{if} \quad \alpha < 1, \end{cases} \tag{5.7}$$

$$0 < \tau_2 = \tau_2(\beta) < \frac{1}{5-2\beta} . \tag{5.8}$$

Then, with

$$A_n(y) = \int_{y \wedge U_n(y)}^{y \vee U_n(y)} B_n(u)\,dQ(u) \quad ,$$

(5.9) $$\Delta_n^{(4)} \le \sup_{0<y\le\varepsilon_{1n}} |A_n(y)| + \sup_{\varepsilon_{1n}\le y\le 1/2} |A_n(y)|$$

$$+ \sup_{1/2\le y\le\varepsilon_{2n}} |A_n(y)| + \sup_{1-\varepsilon_{2n}\le y<1} |A_n(y)| .$$

Using the fact that $\varepsilon_{2n} > 2((\log\log n)/n)^{\frac{1}{2}}$ and the loglog law for $U_n(y)-y$, we see that

(5.10) $$\begin{aligned}(1-\varepsilon_{2n}) \wedge U_n(1-\varepsilon_{2n}) &\ge (1-\varepsilon_{2n}) \wedge \{(1-\varepsilon_{2n})-2((\log\log n)/n)^{\frac{1}{2}}\} \\ &\ge (1-\varepsilon_{2n}) \wedge \{(1-\varepsilon_{2n})-\varepsilon_{2n}\} \\ &= 1-2\varepsilon_{2n}\end{aligned}$$

for all $n \ge n_1 = n_1(\omega)$, where the random variable n_1 is almost surely finite. Similarly,

(5.11) $$\varepsilon_{1n} \vee U_n(\varepsilon_{1n}) \le 2\varepsilon_{1n}$$

for all $n \ge n_1$. Estimating the first term in (5.9) for such n's , we obtain by (5.11), Lemma 2.2, (5.4) and (5.5) that

$$\begin{aligned}\sup_{0<y\le\varepsilon_{1n}} |A_n(y)| &\le \int_0^{2\varepsilon_{1n}} |B_n(u)|\,dQ(u) \\ &\le \sup_{0\le u\le\varepsilon_{1n}} |B_n(u)|\,Q(2\varepsilon_{1n}) \\ &\overset{\text{a.s.}}{=} \begin{cases} O((\varepsilon_{1n}\log\log \varepsilon_{1n})^{\frac{1}{2}}(\log\log n)^{\frac{1}{2}}), & \alpha \ge 1 \\ O(\varepsilon_{1n}^{3/2-\alpha}(\log\log \varepsilon_{1n})^{\frac{1}{2}}(\log\log n)^{\frac{1}{2}}), & \alpha < 1 \end{cases} \\ &= O(n^{-\rho_1}), \quad \rho_1 < \nu(\alpha).\end{aligned}$$

Using also condition (5.2) on $(0,1/2]$ with a resulting new constant C_6 and (2.9) with an arbitrarily small $\delta > 0$, for $n \ge n_1 \vee n_2$, where $n_2 = n_2(\omega)$ is also a.s. finite, we obtain for the second term that

$$\sup_{\varepsilon_{1n}\le y\le 1/2} |A_n(y)| = \sup_{\varepsilon_{1n}\le y\le 1/2} \left| \int_{y\wedge U_n(y)}^{y\vee U_n(y)} \frac{B_n(u)}{f(Q(u))}\,du \right|$$

$$\le C_6 \sup_{0\le y\le 1/2} \frac{B_n(y)}{y^{\frac{1}{2}-\delta}} \sup_{\varepsilon_{1n}\le y\le 1/2} \left| \int_{y\wedge U_n(y)}^{y\vee U_n(y)} u^{\frac{1}{2}-\alpha-\delta} du \right|$$

$$\le 3C_6(\mathrm{loglog}\ n)^{\frac{1}{2}} \sup_{\varepsilon_{1n}\le y\le 1/2} \left| \int_{y\wedge U_n(y)}^{y\vee U_n(y)} \frac{1}{u}\, du \right|$$

$$= 3C_6(\mathrm{loglog}\ n)^{\frac{1}{2}} \sup_{\varepsilon_{1n}\le y\le 1/2} \left| \log\left(1 + \frac{y\vee U_n(y) - y\wedge U_n(y)}{y\wedge U_n(y)}\right) \right|$$

$$\le 3C_6(\mathrm{loglog}\ n)^{\frac{1}{2}} \sup_{\varepsilon_{1n}\le y\le 1/2} \left| \frac{y\vee U_n(y) - y\wedge U_n(y)}{y\wedge U_n(y)} \right|$$

$$\le 6C_6(\mathrm{loglog}\ n)^{\frac{1}{2}} \frac{((\mathrm{loglog}\ n)/n)^{\frac{1}{2}}}{\varepsilon_{1n} - 2((\mathrm{loglog}\ n)/n)^{\frac{1}{2}}}$$

$$\le C_7\, \varepsilon_{1n}^{-1}\, n^{-\frac{1}{2}} \mathrm{loglog}\ n$$

$$= O(n^{-\rho_1}), \quad \rho_1 < \nu(\alpha)\ ,$$

where, in the third step, we used the assumption that $\alpha < 3/2$. Replacing α by β in the above argument and using condition (5.2) on $[1/2,1)$ in the second step, we obtain in exactly the same way that

$$\sup_{1/2\le y\le 1-\varepsilon_{2n}} |A_n(y)| \overset{a.s.}{=} O(\varepsilon_{2n}^{-1}\, n^{-\frac{1}{2}} \mathrm{loglog}\ n)$$

$$= O(n^{-\rho_2}), \quad \rho_2 < \frac{3-2\beta}{10-4\beta}.$$

For the fourth term in (5.9) we get by (5.10) that

$$\sup_{1-\varepsilon_{2n}\le y<1} |A_n(y)| \le \int_{1-2\varepsilon_{2n}}^{1} |B_n(y)|\, dQ(y)$$

$$= I_n^{(4)}(2\varepsilon_{2n})$$

for $n \ge n_1$, and by (3.7) and (5.3) we have

$$I_n^{(4)}(2\varepsilon_{2n}) \overset{a.s.}{=} O(n^{-\tau_2(\frac{1}{2}-\gamma-\delta)})$$

$$= O(n^{-\tau_2(\frac{3}{2}-\beta-\delta')})$$

$$= O(n^{-\rho_2}), \quad \rho_2 < \frac{3-2\beta}{10-4\beta}\ ,$$

since $\delta' > 0$ is arbitrarily small. Collecting the four estimates, the lemma is proved.

In a later section we shall require some exact rates for the increments of B_n itself. Lemma 1.1.1 in M. Csörgő and Révész (1981), proved for a Wiener process, readily implies that for any $\varepsilon > 0$ there exists a constant $C(\varepsilon)$ such that the inequaity

$$P\{\sup_{0\le u\le 1-h}\ \sup_{0\le y\le h} |B_n(u+y)-B_n(y)| \ge vh^{\frac{1}{2}}\} \le \frac{C(\varepsilon)}{h} e^{-\frac{v^2}{2+\varepsilon}}$$

holds for every positive v and $h \in (0,1)$. This inequality, in turn, by the Borel-Cantelli lemma implies the following

LEMMA 5.4. <u>Almost surely,</u>

$$\limsup_{n\to\infty} (\log n)^{-\frac{1}{2}} \sup_{0\le u\le 1-h}\ \sup_{0\le y\le h} |B_n(u+y)-B_n(y)| \le \sqrt{2h}.$$

6. TOTAL TIME ON TEST PROCESSES

Let $Q_n = F_n^{-1}$ be the quantile function of the original sample. Then

$$\int_0^{U_n(y)} (1-E_n(u))\,dQ(u) = \int_0^{Q(U_n(y))} (1-E_n(F(x)))\,dx$$

$$= \int_0^{Q_n(y)} (1-F_n(x))\,dx$$

almost surely by the continuity of F, and the possible exceptional set, where ties may occur among the sample elements, does not depend on y. If $(k-1)/n \leq y < k/n$ for some $k=1,\ldots,n$, then the last integral is

$$\int_0^{X_{k:n}} (1-F_n(x))\,dx = \sum_{i=1}^{k} \int_{X_{i-1:n}}^{X_{i:n}} (1-F_n(x))\,dx$$

$$= \sum_{i=1}^{k} (1 - \frac{i-1}{n})(X_{i:n} - X_{i-1:n})$$

$$= \frac{1}{n} \sum_{i=1}^{k} (n+1-i)(X_{i:n} - X_{i-1:n})$$

$$= \frac{1}{n} \sum_{i=1}^{[ny]+1} W_{i:n}$$

in which we recognise $H_n^{-1}(y)$ of (1.3). Since the above integrals are $\overline{X}_n$ at $y=1$, we arrived at the observation that

$$\text{(6.1)} \qquad P\{\sup_{0\leq y\leq 1} |H_n^{-1}(y) - \int_0^{U_n(y)} (1-E_n(u))\,dQ(u)| = 0\} = 1$$

for each n. This observation will be basic for all the results in this section. Theorem 2.1 of Barlow and van Zwet (1970), designed for a more general transform which contains the total time on test transform as a special case, implies that if $\mu < \infty$ and if F has a continuous density, then $\sup\{|H_n^{-1}(y) - H_F^{-1}(y)| : 0 \leq y \leq F(X_{n:n})\} \xrightarrow{a.s.} 0$. Clearly, the smoothness condition is too strong here. Otherwise, as far as we know, almost sure consistency was proved (Langberg, Leon and Proschan (1980), Theorem 3.2) only pointwise up to now. They also observed that if Q is discontinuous at $y \in (0,1)$, then $H_n^{-1}(y)$ could not converge to $H_F^{-1}(y)$ almost surely. We have

THEOREM 6.1. _If_ $\mu < \infty$ _and_ $Q = F^{-1}$ _is continuous on_ $[0,1)$, then

$$\Delta_n^{(5)} = \sup_{0\le y\le 1} |H_n^{-1}(y) - H_F^{-1}(y)| \xrightarrow{a.s.} 0 .$$

Proof. By (6.1)

$$\Delta_n^{(5)} \overset{a.s.}{=} \sup_{0\le y\le 1} \left| \int_0^{U_n(y)} (1-E_n(u))\,dQ(u) - \int_0^y (1-u)\,dQ(u) \right|$$

$$\le \sup_{0\le y\le 1} \left| \int_0^{U_n(y)} (1-E_n(u))\,dQ(u) - \int_0^{U_n(y)} (1-u)\,dQ(u) \right|$$

$$+ \sup_{0\le y\le 1} \left| \int_0^{U_n(y)} (1-u)\,dQ(u) - \int_0^y (1-u)\,dQ(u) \right|$$

$$\le \Delta_n^{(1)} + \sup_{0\le y\le 1} |H_F^{-1}(U_n(y)) - H_F^{-1}(y)| ,$$

where $\Delta_n^{(1)} \xrightarrow{a.s.} 0$ by Lemma 3.1. Since $U_n(y)$ converges a.s. uniformly to y, and the function H_F^{-1} is uniformly continuous on $[0,1]$, the theorem follows.

The following sufficient conditions for weak convergence are close to being optimal. This fact will be demonstrated after the proof.

THEOREM 6.2. Suppose that the density function $f = F'$ is continuous and positive on the open support of F. If

$$J = \sup_{0<u<1} \frac{q(u)(1-u)}{f(Q(u))} < \infty \tag{6.2}$$

for some O'Reilly weight function q and $EX^2 < \infty$, then

$$\Delta_n^{(6)} = \sup_{0\le u\le 1} |t_n(u) - T_n(u)| \xrightarrow{P} 0$$

with the Gaussian processes

$$T_n(u) = \int_0^u B_n(y)\,dQ(y) + \frac{1-u}{f(Q(u))} B_n(u), \quad 0 \le u \le 1. \tag{6.3}$$

Proof. By the basic observation (6.1) and the notation of (3.1) we have

$$t_n(y) - T_n(y) = \beta_n(U_n(y)) - \int_0^u B_n(u)\,dQ(u)$$

$$+ n^{1/2}\{H_F^{-1}(U_n(y)) - H_F^{-1}(y)\} - \frac{1-y}{f(Q(y))} B_n(y)$$

$$= \beta_n(U_n(y)) - \int_0^{U_n(y)} B_n(u)\,dQ(u)$$

$$+ \int_0^{U_n(y)} B_n(u)\,dQ(u) - \int_0^{y} B_n(u)\,dQ(u)$$

$$+ n^{\frac{1}{2}}\{H_F^{-1}(U_n(y)) - H_F^{-1}(y)\} - \frac{1-y}{f(Q(y))} B_n(y)$$

almost surely for each fixed n. Hence

$$\Delta_n^{(6)} \leq \Delta_n^{(2)} + \Delta_n^{(4)} + \Delta_n^{(7)},$$

where $\Delta_n^{(2)}$ of Lemma 3.2 and $\Delta_n^{(4)}$ of Lemma 5.2 converge to zero in probability only under $EX^2 < \infty$ and the continuity of Q. What remains to show now is that

$$(6.4) \qquad \Delta_n^{(7)} = \sup_{0 \leq y \leq 1} |n^{\frac{1}{2}}\{H_F^{-1}(U_n(y)) - H_F^{-1}(y)\} - \frac{1-y}{f(Q(y))} B_n(y)|$$

converges to zero. Let $\eta \in (0,1)$ and $\delta > 0$ be given. By Remark 1 of Wellner (1978), as formulated in Lemma 2.7, there exists a $\lambda = \lambda(\eta) \geq 1$ such that

$$(6.5) \qquad P\{\frac{y}{\lambda} \leq U_n(y),\ \frac{1}{n} \leq y < 1\} > 1-\eta$$

for all n. For this λ and for an arbitrary positive constant C_9 we have

$$(6.6) \qquad \lim_{\varepsilon \to 0} \limsup_{n \to \infty} P\{\sup_{\frac{1}{n} \leq y \leq \varepsilon} \frac{|u_n(y)|}{q(y/\lambda)} > C_9\} = 0,$$

because, evidently, $q(y/\lambda)$ is also an O'Reilly weight function if q is such. Also, by (2.18),

$$(6.7) \qquad \lim_{\varepsilon \to 0} P\{\sup_{0 \leq y \leq \varepsilon} \frac{|B_n(y)|}{q(y)} > C_{10}\} = 0$$

for any positive constant C_{10}. Then by (6.6) and (6.7) we can choose $\varepsilon \in (0,1/2)$ so small that

$$(6.8) \qquad \limsup_{n \to \infty} P\{\sup_{1/n \leq y \leq \varepsilon} \frac{|u_n(y)|}{q(y/\lambda)} > \frac{\delta}{2J}\} \leq \eta,$$

and

$$(6.9) \qquad P\{\sup_{0<y\le\varepsilon} \frac{|B_n(y)|}{q(y)} > \frac{\delta}{J}\} \le \eta, \quad n=1,2,\ldots .$$

For $n > 1/\varepsilon$ we have

$$\begin{aligned}
\Delta_n^{(7)} \le & \sup_{0\le y\le 1/n} |n^{\frac{1}{2}}\{H_F^{-1}(U_n(y))-H_F^{-1}(y)\}| \\
& + \sup_{1/n\le y\le\varepsilon} |n^{\frac{1}{2}}\{H_F^{-1}(U_n(y))-H_F^{-1}(y)\}| \\
& + \sup_{0<y\le\varepsilon} |\frac{1-y}{f(Q(y))} B_n(y)| \\
& + \sup_{\varepsilon\le y\le 1-\varepsilon} |n^{\frac{1}{2}}\{H_F^{-1}(U_n(y))-H_F^{-1}(y)\} - \frac{1-y}{f(Q(y))} B_n(y)| \qquad (6.10) \\
& + \sup_{1-\varepsilon\le y<1} |\frac{1-y}{f(Q(y))} B_n(y)| \\
& + \sup_{1-\varepsilon\le y\le(n-1)/n} |n^{\frac{1}{2}}\{H_F^{-1}(U_n(y))-H_F^{-1}(y)\}| \\
& + \sup_{(n-1)/n\le y\le 1} |n^{\frac{1}{2}}\{H_F^{-1}(U_n(y))-H_F^{-1}(y)\}| \\
= & A_n^{(1)} + A_n^{(2)}(\varepsilon) + A_n^{(3)}(\varepsilon) + A_n^{(4)}(\varepsilon,1-\varepsilon) + A_n^{(5)}(\varepsilon) \\
& + A_n^{(6)}(\varepsilon) + A_n^{(7)} .
\end{aligned}$$

Applying a one-term Taylor expansion and the fact that $(H_F^{-1}(y))' = (1-y)/f(Q(y))$, for the middle term we obtain

$$\begin{aligned}
A_n^{(4)}(\varepsilon,1-\varepsilon) \le & \sup_{\varepsilon\le y\le 1-\varepsilon} \left|\frac{1-\tau_n(y)}{f(Q(\tau_n(y)))} - \frac{1-y}{f(Q(y))}\right| \sup_{\varepsilon\le y\le 1-\varepsilon} |u_n(y)|, \\
\le & \sup_{\varepsilon\le y\le 1-\varepsilon} \frac{1}{f(Q(y))} \sup_{\varepsilon\le y\le 1-\varepsilon} |u_n(y)-B_n(y)| \\
& + \sup_{\varepsilon\le y\le 1-\varepsilon} \left|\frac{1-\tau_n(y)}{f(Q(\tau_n(y)))} - \frac{1-y}{f(Q(y))}\right| \sup_{\varepsilon\le y\le 1-\varepsilon} |u_n(y)|,
\end{aligned}$$

where

$$(6.11) \qquad y \wedge U_n(y) \le \tau_n(y) \le y \vee U_n(y).$$

The latter relation implies that $\tau_n(y)$ converges a.s. uniformly to y on [0,1]. Since the compound function $f(Q(\cdot))$ is uniformly continuous on $[\varepsilon,1-\varepsilon]$ by assumption, and since $\sup|u_n(y)|$ has a limit distribution, the second term converges to zero a.s.. The first term goes to zero in probability by Lemma 2.1.

Now we consider $A_n^{(1)}$. Consider (6.2) implies that for $0 \leq u \leq 1/2$

$$\frac{1}{f(Q(u))} \leq \frac{C_{11}}{u^{\frac{1}{2}}h(u)}$$

with the representation of q in (2.24), where $h(u) \to\infty$ as $u \to 0$ by Lemma 2.6. Hence, for $0 < y \leq 1/2$,

$$
\begin{aligned}
Q(y) &= \int_0^y Q'(u)\,du \\
&\leq C_{11} \int_0^y \frac{1}{u^{\frac{1}{2}}h(u)}\,du \\
&\leq C_{11} \frac{1}{\hat{h}(y)} \int_0^y u^{-\frac{1}{2}}du \qquad (6.12)\\
&= 2C_{11} \frac{1}{\hat{h}(y)}\, y^{\frac{1}{2}},
\end{aligned}
$$

where $\hat{h}(y) = \inf\{h(u) : 0 < u < y\} \to \infty$ as $y \to 0$. Therefore

$$
\begin{aligned}
A_n^{(1)} &\leq n^{\frac{1}{2}} H_F^{-1}(U_{1:n}) + n^{\frac{1}{2}} H_F^{-1}(\tfrac{1}{n}) \\
&= n^{\frac{1}{2}} \int_0^{U_{1:n}} (1-v)\,dQ(v) + n^{\frac{1}{2}} \int_0^{1/n} (1-v)\,dQ(v) \\
&\leq 2C_{11}(nU_{1:n})^{\frac{1}{2}}(\hat{h}(U_{1:n}))^{-1} + 2C_{11}(\hat{h}(\tfrac{1}{n}))^{-1},
\end{aligned}
$$

and this bound goes to zero in probability since $(\hat{h}(U_{1:n}))^{-1} \xrightarrow{P} 0$, and $(nU_{1:n})^{\frac{1}{2}}$ has a limiting distribution (apparently a Weibull (2)).

Next we consider $A_n^{(2)}(\varepsilon)$ in (6.10). Clearly, by (6.2) and the monotonicity of q,

$$
\begin{aligned}
A_n^{(2)}(\varepsilon) &= \sup_{\frac{1}{n} \leq y < \varepsilon} n^{\frac{1}{2}} \left| \int_0^{U_n(y)} \frac{u}{f(Q(u))} du \right| \\
&\leq J \sup_{\frac{1}{n} \leq y \leq \varepsilon} n^{\frac{1}{2}} \left| \int_y^{U_n(y)} \frac{1}{q(u)}\, du \right| \\
&\leq J \left\{ \sup_{\frac{1}{n} \leq y \leq \varepsilon} \frac{|u_n(y)|}{q(y)} + \sup_{\frac{1}{n} \leq y \leq \varepsilon} \frac{|u_n(y)|}{q(U_n(y))} \right\}.
\end{aligned}
$$

Thus, still by the monotonicity of q and by (6.5),

$$P\{A_n^{(2)} \geq \delta\} \leq \eta + P\{\sup_{\frac{1}{n} \leq y \leq \varepsilon} \frac{|u_n(y)|}{q(y/\lambda)} > \frac{\delta}{2J}\},$$

whence by (6.8),

$$\limsup_{n\to\infty} P\{A_n^{(2)}(\varepsilon) > \delta\} \leq 2\eta.$$

The third term is easy. Clearly,

$$A_n^{(3)}(\varepsilon) \leq J \sup_{0<y\leq\varepsilon} \frac{|B_n(y)|}{q(y)},$$

and (6.9) takes care of this bound. Summing up, presently we have

$$\limsup_{n\to\infty} P\{\sup_{0\leq y<1-\varepsilon} |n^{\frac{1}{2}}\{H_F^{-1}(U_n(y))-H_F^{-1}(y)\} - \frac{1-y}{f(Q(y))} B_n(y)| > 2\delta\} \leq 3\eta.$$

Turning now to $A_n^{(7)}$, we first note that condition (6.2) implies that for $1/2 \leq u < 1$

$$\frac{1-u}{f(Q(u))} \leq \frac{C_{12}}{(1-u)^{\frac{1}{2}}h(1-u)}.$$

Hence, for $1/2 \leq y < 1$ and with the above $\hat{h}(1-y) = \inf\{h(1-u) : y \leq u < 1\}$,

$$\int_y^1 \frac{1-u}{f(Q(u))}\, du \leq C_{12} \frac{1}{\hat{h}(1-y)} \int_y^1 (1-u)^{-\frac{1}{2}} du$$

$$\leq 2C_{12} \frac{1}{\hat{h}(1-y)} (1-y)^{\frac{1}{2}}.$$

Therefore,

$$A_n^{(7)} \leq n^{\frac{1}{2}} \int_{U_{n:n}}^1 (1-u)\,dQ(u) + n^{\frac{1}{2}} \int_{1-\frac{1}{n}}^1 (1-u)\,dQ(u)$$

$$\leq 2C_{12}\{(u(1-U_{n:n}))^{\frac{1}{2}}(\hat{h}(1-U_{n:n}))^{-1} + (\hat{h}(\tfrac{1}{n}))^{-1}\}$$

and this converges again to zero in probability just as the bound for $A_n^{(1)}$ did.

The estimation techniques for handling $A_n^{(2)}(\varepsilon)$ and $A_n^{(3)}(\varepsilon)$ transfer now to handling $A_n^{(5)}(\varepsilon)$ and $A_n^{(6)}(\varepsilon)$, respectively, if we modify them in a similar way as the technique for $A_n^{(1)}$ was modified when treating $A_n^{(7)}$. In this way we obtain that

$$\limsup_{n\to\infty} P\{\Delta_n^{(7)} > 4\delta\} \leq 6\eta,$$

and hence the theorem.

In order to verify our claim on the optimality of the conditions of Theorem 6.2, we first note that condition (6.2) is slightly stronger than $J(2,2/3) < \infty$, on recalling the notation in (2.24) and (5.2). Hence (5.3) implies that $J(r) < \infty$ for any $r < 2$. On the other hand, it does not imply that $J(2) < \infty$, and not even that $EX^2 < \infty$, and this is why we assumed the latter condition separately. If we assume the somewhat stronger condition

$$J\left(\frac{2}{1-\delta}, \frac{2}{3-\delta}\right) = \sup_{0<u<1} \frac{u^{(1-\delta)/2}(1-u)^{(3-\delta)/2}}{f(Q(u))} < \infty, \quad 0 < \delta < 1,$$

instead of (6.2), i.e., if we prescribe the h function in (2.24) to be $h(u) = u^{-\delta/2}$, $0 < u \le 1/2$, in which case condition (6.2) is of course automatically satisfied, then $J(r) < \infty$ with any $r \in [2, 2/(1-\delta))$, and hence $EX^2 < \infty$. Suppose now that $EX^2 < \infty$, but almost to the contrary of (6.2),

$$\sup_{0<u<1} \frac{q(u)(1-u)}{f(Q(u))} = \infty$$

for a function $q \in Q^*$ for which the integral in (2.5) diverges. Then, by (3.4) or Lemma 5.1, the first term of our limit process

$$T(u) = \int_0^u B(y)\,dQ(y) - \frac{1-u}{f(Q(u))} B(u)$$

is almost surely bounded, but the second term is almost surely unbounded on [0,1]. Hence $T(\cdot)$ is almost surely unbounded and thus no process can converge to it weakly in the space of continuous functions on [0,1]. This is why condition (6.2) is almost optimal. For further discussion of this condition we refer to Section 8.

THEOREM 6.3. _Suppose that the density function_ $f = F'$ _is positive on the open support of_ F, _and_

(6.13) $$J\left(\frac{1}{\alpha}, \frac{1}{\beta}\right) = \sup_{0<u<1} \frac{u^{\alpha}(1-u)^{\beta}}{f(Q(u))} < \infty \text{ with } 0 \le \alpha < \frac{1}{8^{1/2}}, \quad 0 \le \beta < 1 + \frac{1}{8^{1/2}}.$$

If, moreover,

(6.14) $$J_1 = \sup_{0<u<1} u(1-u) \frac{|f'(Q(u))|}{f^2(Q(u))} < \infty$$

holds for the derivative $f' = F''$ _of the density function, then_

$$\Delta_n^{(6)} = \sup_{0<u<1} |t_n(u) - T_n(u)| \overset{a.s.}{=} O(n^{-\tau})$$

for any $\tau < \begin{cases} \min(\frac{1}{4}, \frac{1-8\alpha^2}{8\alpha+4}, \frac{3-2\beta}{10-4\beta}), & \text{if } 0 \le \beta \le 1, \\ \min(\frac{1-8\alpha^2}{8\alpha+4}, \frac{3-2\beta}{10-4\beta}, \frac{1-8(\beta-1)^2}{8(\beta-1)^2+4}), & \text{if } 1 < \beta < 1+8^{-1/2}. \end{cases}$

Proof. Just as in the preceding proof, $\Delta_n^{(6)} \le \Delta_n^{(2)} + \Delta_n^{(4)} + \Delta_n^{(7)}$, where

$$\Delta_n^{(2)} \overset{a.s.}{=} O(n^{-\lambda}), \quad \lambda < \min(\tfrac{1}{2}, \tfrac{3}{2} - \beta), \tag{6.15}$$

by Lemma 3.3 and the implication in (5.3), and

$$\Delta_n^{(4)} \overset{a.s.}{=} O(n^{-\rho}), \quad \rho < \min\{\tfrac{3-2\alpha}{10-4\alpha}, \tfrac{3-2\beta}{10-4\beta}\} \tag{6.16}$$

by Lemma 5.3. Now it remains only to estimate $\Delta_n^{(7)}$ of (6.4). Set $\varepsilon_{1n} = n^{-\varepsilon_1}$, where

$$0 < \varepsilon_1 < \frac{4\alpha+1}{4\alpha+2},$$

and let

$$\varepsilon_{2n} = \begin{cases} 25n^{-1} \log\log n, & \text{if } 0 < \beta \le 1, \\ n^{-\varepsilon_2}, \ 0 < \varepsilon_2 < \frac{4(\beta-1)+1}{4(\beta-2)+2}, & \text{if } 1 < \beta < 8^{-1/2} + 1. \end{cases}$$

We break up $\Delta_n^{(7)}$ according to (6.10), but with the ε replaced by ε_{1n} and ε_{2n}:

$$\Delta_n^{(7)} \le A_n^{(1)} + A_n^{(2)}(\varepsilon_{1n}) + A_n^{(3)}(\varepsilon_{1n}) + A_n^{(4)}(\varepsilon_{1n}, 1-\varepsilon_{2n}) + A_n^{(5)}(\varepsilon_{2n}) + A_n^{(6)}(\varepsilon_{2n}) + A_n^{(7)}.$$

Consider $A_n^{(1)}$. Condition (6.13) implies that for $0 < u \le 1/2$,

$$\frac{1}{f(Q(u))} \le C_{13} u^{-\alpha}. \tag{6.17}$$

Proceeding similarly as in the proof of Theorem 6.2, we obtain

$$A_n^{(1)} \le (1-\alpha)^{-1} C_{13} (nU_{1:n})^{1-\alpha} n^{\alpha-1/2} + (1-\alpha)^{-1} C_{13} n^{\alpha-1/2}.$$

Since $nU_{1:n} = O(\log n)$ a.s. by Lemma 2.10, we get

$$A_n^{(1)} \overset{a.s.}{=} O(n^{\alpha-1/2}(\log n)^{1-\alpha}). \tag{6.18}$$

Next, by (6.17),

$$A_n^{(2)}(\varepsilon_{1n}) = \sup_{\frac{1}{n} \le y \le \varepsilon_{1n}} n^{1/2} \left| \int_y^{U_n(y)} (1-u)\, dQ(u) \right|$$

$$\le C_{13} \sup_{\frac{1}{n} \le y \le \varepsilon_{1n}} n^{1/2} \left| \int_y^{U_n(u)} u^{-\alpha} du \right|$$

$$\leq C_{13}' \sup_{\frac{1}{n} \leq y \leq \varepsilon_{1n}} \max(y^{-\alpha}, (U_n(y))^{-\alpha}) |u_n(y)|$$

$$\leq C_{13} \max(n^{\alpha}, U_{1:n}^{-\alpha}) \sup_{0 \leq y \leq \varepsilon_{1n}} |u_n(y)|.$$

By Lemma 2.10 we have

$$U_{1:n}^{-\alpha} \stackrel{a.s.}{=} O(n^{\alpha}(\log n)^{2/\alpha}) ,$$

and using (2.12) and the definition of ε_{1n}, we have

$$\sup_{0 \leq y \leq \varepsilon_{1n}} |u_n(y)| \stackrel{a.s.}{=} O(\varepsilon_{1n}^{1/2}(\log n)^{1/2}).$$

Hence

(6.19) $$A_n^{(2)}(\varepsilon_{1n}) \stackrel{a.s.}{=} O(n^{\alpha}\varepsilon_{1n}^{1/2}(\log n)^{2/\alpha+1/2}).$$

Applying (6.17) again and (2.9), we obtain for the third term that

(6.20) $$A_n^{(3)}(\varepsilon_{1n}) \stackrel{a.s.}{=} O(\varepsilon_{1n}^{(1/2-\alpha-\delta)}(\log\log n)^{1/2})$$

with an arbitrarily small $\delta > 0$.

To estimate the fourth term, we first note that by simple derivation

$$(H_F^{-1})''(u) = \frac{d^2}{du^2} H_F^{-1}(u) = -\frac{1}{f(Q(u))} - \frac{(1-u)f'(Q(u))}{f^2(Q(u))} \frac{1}{f(Q(u))},$$

and hence by condition (6.14),

(6.21) $$|(H_F^{-1})''(u)| \leq (1 + \frac{J_1}{u}) \frac{1}{f(Q(u))}.$$

The second derivative comes in when applying a two-term Taylor expansion:

(6.22) $$n^{1/2}\{H_F^{-1}(U_n(y)) - H_F^{-1}(y)\}$$

$$= \frac{1-y}{f(Q(y))} u_n(y) + (H_F^{-1})''(\tau_n(y))n^{-1/2}u_n^2(y),$$

where $\tau_n(y)$ satisfies the inequalities in (6.11). We shall break up the fourth term into two terms

$$A_n^{(4)}(\varepsilon_{1n}, 1-\varepsilon_{2n}) \leq A_n^{(4)}(\varepsilon_{1n}, \tfrac{1}{2}) + A_n^{(4)}(\tfrac{1}{2}, 1-\varepsilon_{2n}) ,$$

and first we estimate the first term on the right hand side.

(6.23) $$A_n^{(4)}(\varepsilon_{1n}, \tfrac{1}{2}) \leq \sup_{\varepsilon_{1n} \leq y \leq \frac{1}{2}} \frac{1-y}{f(Q(y))} |u_n(y) - B_n(y)|$$

$$+ \sup_{\varepsilon_{1n} \leq y \leq \frac{1}{2}} \left(1 + \frac{J_1}{\tau_n(y)}\right) \frac{1}{f(Q(\tau_n(y)))} n^{-1/2} u_n^2(y)$$

by (6.21), and the first term here is almost surely

$$O(\varepsilon_{1n}^{\alpha}\, n^{-1/4}(\log\log n)^{1/4}(\log n)^{1/2}).$$

But (6.11) and Lemma 2.9 imply that

$$(6.24)\qquad \sup_{\varepsilon_{1n}\le y\le\frac{1}{2}} \frac{y}{\tau_n(y)} \le 6 \quad \text{a.s.}$$

for large enough (random) n, provided $\varepsilon_{1n} \ge \delta_n = 25n^{-1}\log\log n$ (which inequality holds here), while (2.25) states that

$$(6.25)\qquad \sup_{\delta_n\le y\le 1-\delta_n} \frac{u_n^2(y)}{y(1-y)} \overset{\text{a.s.}}{=} O(\log\log n).$$

Hence the second term in (6.22) is not greater, on applying also condition (6.19), than

$$C_{14}n^{-\frac{1}{2}} \sup_{\varepsilon_{1n}\le y\le\frac{1}{2}} y^{-\alpha}\left(\frac{y}{\tau_n(y)}\right)^{1+\alpha} \frac{\tau_n^{\alpha}(y)}{f(Q(\tau_n(y)))}\,\frac{u_n^2(y)}{y}$$

$$\overset{\text{a.s.}}{=} O(n^{-\frac{1}{2}}\,\varepsilon_{1n}^{\alpha}\,\log\log n).$$

Introducing now the notation

$$A_n(y) = n^{\frac{1}{2}}\{H_F^{-1}(U_n(y))-H_F^{-1}(y)\} - \frac{1-y}{f(Q(y))}\,B_n(y)$$

for the process in $\Delta_n^{(7)}$ of (6.4), we have

$$(6.26)\qquad \Delta_n^{(7)} \le \sup_{0\le y\le 1/2} |A_n(y)| + \sup_{1/2\le y\le 1} |A_n(y)|,$$

where

$$(6.27)\qquad \sup_{0\le y\le 1/2} |A_n(y)| \le A_n^{(4)}(\tfrac{1}{2},1-\varepsilon_{2n})+A_n^{(5)}(\varepsilon_{2n})+A_n^{(6)}(\varepsilon_{2n})+A_n^{(7)},$$

and the last two order relations, (6.23), (6.18), (6.19) and (6.20) together with the definition of ε_{1n} give by elementary computations that

$$(6.28)\qquad \sup_{0\le y\le 1/2} |A_n(y)| \overset{\text{a.s.}}{=} O(n^{-\tau_1}),\quad \tau_1 < \frac{8\alpha^2-1}{8\alpha+4}.$$

The estimation of $A_n^{(7)}$ is completely analogous to that of $A_n^{(1)}$, and we obtain

$$(6.29)\qquad A_n^{(7)} \overset{\text{a.s.}}{=} O(n^{\beta-\frac{3}{2}}(\log n)^{2-\beta}).$$

Here we used that

$$(6.30)\qquad \frac{1-u}{f(Q(u))} \le C_{15}(1-u)^{-(\beta-1)},\quad \frac{1}{2} \le u < 1,$$

analogously to (6.17), and using the last inequality when estimating $A_n^{(6)}(\varepsilon_{2n})$ we obtain

$$A_n^{(6)}(\varepsilon_{2n}) \le C_{15} \sup_{1-\varepsilon_{2n}\le y\le \frac{n-1}{n}} n^{\frac{1}{2}} \left|\int_y^{U_n(y)} (1-u)^{1-\beta} du\right|.$$

Now the cases $0 < \beta \le 1$ and $1 < \beta < 1+8^{-\frac{1}{2}}$ should be separated. In the first one, by (2.13) and the definition of ε_{2n}, we get

(6.31)
$$A_n^{(6)}(\varepsilon_{2n}) \le C_{16} \sup_{1-\varepsilon_{2n}\le y\le 1} |u_n(y)| \overset{a.s.}{=} O(n^{-\frac{1}{2}} \log n),$$

while if $1 < \beta < 1+8^{-\frac{1}{2}}$, then

(6.32)
$$A_n^{(6)}(\varepsilon_{2n}) \le C_{15} \max(n^{\beta-1}, (1-U_{n:n})^{1-\beta}) \sup_{1-\varepsilon_{2n}\le y\le 1} |u_n(y)| \overset{a.s.}{=} O(n^{\beta-1} \varepsilon_{2n}^{\frac{1}{2}} (\log n)^{1/2 + 2/\beta-1})$$

by the argument that led to (6.19).

By (6.30) and (2.9), with an arbitrarily small $\delta > 0$,

(6.33)
$$A_n^{(5)}(\varepsilon_{2n}) \overset{a.s.}{=} O(\varepsilon_{2n}^{(3/2-\beta-\delta)} (\log\log n)^{\frac{1}{2}}).$$

Finally, the estimation of $A_n^{(4)}(\frac{1}{2}, 1-\varepsilon_{2n})$ is analogous to that of $A_n^{(4)}(\varepsilon_{1n}, \frac{1}{2})$, but we have to separate the above two cases again. If $0 < \beta \le 1$, then $(1-y)/f(Q(y))$ and $(1-y)^{-(\beta-1)}$ are bounded on $[1/2, 1]$, and thus we obtain

$$A_n^{(4)}(\tfrac{1}{2}, 1-\varepsilon_{2n}) \overset{a.s.}{=} \begin{cases} O(n^{-1/4} (\log\log n)^{1/4} (\log n)^{1/2}), & \text{if } 0 < \beta \le 1, \\ O(\varepsilon_{2n}^{\beta-1} n^{-1/4} (\log\log n)^{1/4} (\log n)^{1/2}), & \text{if } 1 < \beta < 1+8^{-\frac{1}{2}}. \end{cases}$$

Collecting now (6.29), (6.31), (6.32), (6.33) and the last equation,

(6.34)
$$\sup_{1/2\le y\le 1} |A_n(y)| = (n^{-\tau_2}),$$

where

$$\tau_2 < \begin{cases} \frac{1}{4}, & \text{if } 0 < \beta \le 1, \\ \frac{1-8(\beta-1)^2}{8(\beta-1)+4}, & \text{if } 1 < \beta < 1+8^{-\frac{1}{2}}. \end{cases}$$

Now (6.15), (6.16), (6.26), (6.28) and (6.34) give the theorem via an elementary comparison of the rate sequences.

It is interesting to point out that although we did not use the strong approximation theorem of M. Csörgő and Révész (1978, 1981) for the general quantile process, condition (6.14), under which the latter approximation holds, entered rather naturally.

Fluctuation type results, such as the one in (4.3) for the mean residual life processes, can also be deduced from Theorem 6.1 for $t_n(u)$, but here we formulate only the loglog law for the sake of later reference. The constant on the right side is inherited from the corresponding law for the approximating processes T_n.

COROLLARY 6.4. *Under the conditions of Theorem* 6.3

$$\limsup_{n\to\infty} \left(\frac{n}{\log\log n}\right)^{\frac{1}{2}} \sup_{0\le u\le 1} |H_n^{-1}(u)-H_F^{-1}(u)| \le T(F)$$

almost surely, where

$$T(F) = 2\left\{\int_0^1 h(y)\,dQ(y) + \sup_{0\le y\le 1} \frac{h(y)(1-y)}{f(Q(y))}\right\}$$

with $h(y) = \left(y(1-y)\log\log \frac{1}{y(1-y)}\right)^{\frac{1}{2}}$ $(0 < y < 1)$.

7. SCALED TOTAL TIME ON TEST PROCESSES

Recalling the definitions right below (1.12) of the quantities in question, Theorem 6.1 and the strong law of large numbers readily imply the following consistency result.

THEOREM 7.1. *If* $\mu < \infty$ *and* $Q = F^{-1}$ *is continuous on* $[0,1)$, *then*

$$\sup_{0 \le y \le 1} |D_n^{-1}(y) - D_F^{-1}(y)| \xrightarrow{a.s.} 0.$$

Introduce now the following sequence of mean-zero Gaussian processes, the members of which are the same in distribution for each n:

$$S_n(y) = \mu^{-1} T_n(y) - \mu^{-2} H_F^{-1}(y) T_n(1), \quad 0 \le y \le 1,$$

where $T_n(\cdot)$ is of Theorem 6.2. Recall also the notation in (1.12).

THEOREM 7.2. *If the conditions of Theorem* 6.2 *are satisfied, then*

$$\sup_{0 \le y \le 1} |s_n(y) - S_n(y)| \xrightarrow{P} 0.$$

Proof. By elementary computation

$$\begin{aligned} s_n(y) &= \mu^{-1} t_n(y) - \mu^{-2} H_F^{-1}(y) t_n(1) \\ &\quad + t_n(y) \left\{ \frac{1}{H_n^{-1}(1)} - \frac{1}{H_F^{-1}(1)} \right\} \\ &\quad + \frac{H_F^{-1}(y)}{\mu} t_n(1) \left\{ \frac{1}{H_F^{-1}(1)} - \frac{1}{H_n^{-1}(1)} \right\}, \end{aligned}$$

and the result follows from Theorems 6.1 and 6.2.

THEOREM 7.3. *If the conditions of Theorem 6.3 are satisfied, then*

$$\sup_{0 \le y \le 1} |s_n(y) - S_n(y)| \overset{a.s.}{=} O(n^{-\tau})$$

with the same τ *as in Theorem 6.3.*

Proof. The third term in the above form of s_n is a.s. $O((\log\log n)/n^{\frac{1}{2}})$ by the loglog law of Corollary 6.4 to Theorem 6.3, while the fourth term is a.s. $O(((\log\log n)/n)^{\frac{1}{2}})$ by the same reason, and hence the result is implied by Theorem 6.3.

COROLLARY 7.4. *Under the conditions of Theorem* 7.3

$$\limsup_{n \to \infty} \left(\frac{n}{\log\log n}\right)^{\frac{1}{2}} \sup_{0 \le y \le 1} |D_n^{-1}(y) - D_F^{-1}(y)| \le \frac{2}{\mu} T(F)$$

almost surely, where $T(F)$ *is as in Corollary 6.4.*

We must point out that assuming the continuity of f and $EX^2 < \infty$, Barlow and Proschan (1977) proved that $|s_n(y) - S_n(y)| \xrightarrow{P} 0$ at each fixed $y \in (0,1)$ on an appropriate probability space. Of course, the weak convergence of $s_n(\cdot)$ to $S_F(\cdot) = S_1(\cdot)$ does not follow from such a statement, as it does from Theorem 7.2. Barlow and Proschan (1977) also noted that if the underlying distribution is exponential, $F(x) = 1 - \exp(-x/\mu)$, $x \geq 0$, then a tedious computation of the covariance function of $S_F(\cdot)$ gives that $S_F(\cdot)$ is a Brownian bridge on $[0,1]$. In this respect see also Point 5) in the next section.

8. DISCUSSION OF RESULTS ON TOTAL TIME ON TEST PROCESSES

1) Returning to the discussion of condition (6.2) of Theorem 6.2, let the density function $f = F'$ be continuous and positive on the open support (t_F, T_F) of F, $0 \le t_F < T_F \le \infty$, and define the failure or hazard rate function r of F by

$$r(x) = r_F(x) = \frac{f(x)}{1-F(x)} .$$

This function is then positive and continuous on (t_F, T_F). Assume that r satisfies one of the conditions:

A_1: $r(t_F) > 0$

A_2: $r(t_F) = 0$, and r is nondecreasing on a right-side neighbourhood of t_F,

and also one of the conditions:

B_1: $r(T_F) > 0$

B_2: $r(T_F) = 0$, and r is nonincreasing on a left-side neighbourhood of T_F.

If we also assume that

$$\begin{aligned} E(r(X))^{-2} &= \int_0^\infty \frac{1-F(x)}{r(x)}\, dx \\ &= \int_0^\infty \left(\frac{1-F(x)}{f(x)}\right)^2 dF(x) \\ &= \int_0^1 \left(\frac{1-u}{f(Q(u))}\right)^2 du \\ &= \int_0^1 \left(\frac{1}{r(Q(u))}\right)^2 du < \infty , \end{aligned}$$

then $\ell(t) = 1/r(Q(t))$ satisfies all the conditions of Lemma 2.4. Hence there exists an O'Reilly weight function q such that

$$J = \sup_{0<u<1} \frac{q(u)}{r(Q(u))} = \sup_{0<x<\infty} \frac{q(F(x))}{r(x)} < \infty$$

and this is exactly condition (6.2). This q here may be chosen as $q(u) = (u(1-u))^{1/(2+\delta)}$, $\delta > 0$.

2) The weak limit of our empirical total time on test process $t_n(u)$, $0 \le u \le 1$, is the mean-zero Gaussian process

$$T(u) = T_F(u) = \int_0^u B(y)dQ(y) - \frac{B(u)}{r(Q(u))}$$

$$= \int_0^{Q(u)} B(F(x))dx - \frac{B(u)}{r(Q(u))}, \quad 0 \leq u \leq 1,$$

under the conditions of Theorem 6.2. Let θ be a nonpositive shift parameter and $\sigma > 0$ be a scale parameter. Define $F_{\theta,\sigma}(x) = F((x-\theta)/\sigma)$. Then it is easy to check that

$$T_{F_{\theta,\sigma}}(u) \equiv \sigma T_F(u), \quad 0 \leq u \leq 1. \tag{8.1}$$

This is of no surprise in view of the fact that if $\tilde{t}_n(u)$ denotes the nth total time on test process belonging to $(X_1-\theta)/\sigma,\ldots,(X_n-\theta)/\sigma$, then $\tilde{t}_n(u) \equiv \sigma t_n(u)$.

3) The covariance function of the limit process is obtained by a straightforward but very lengthy and tedious computation. For $s \leq t$,

$$\begin{aligned}\sigma_1(s,t) &= ET_F(s)T_F(t)\\ &= (H_F^{-1}(s))^2 + (Q(s)-H_F^{-1}(s))\{H_F^{-1}(t) + \frac{(1-t)^2}{f(Q(t))} + \frac{(1-s)^2}{f(Q(s))}\}\\ &\quad + \frac{s(1-s)}{f(Q(s))}\{H_F^{-1}(t) - H_F^{-1}(s)\}\\ &= (H_F^{-1}(s))^2 + (Q(s)-H_F^{-1}(s))\{H_F^{-1}(t) + \frac{1-t}{r(Q(t))} + \frac{1-s}{r(Q(s))}\}\\ &\quad + \frac{s}{r(Q(s))}\{H_F^{-1}(t) - H_F^{-1}(s)\}.\end{aligned}$$

Using another form for the integral in $\sigma_1(s,t)$ and the density-quantile function rather than the failure rate, the variance function is

$$\begin{aligned}\sigma_1^2(t) &= \sigma_1(t,t) = (H_F^{-1}(t))^2 + (Q(t))-H_F^{-1}(t))\ \{H_F^{-1}(t)\\ &\quad + 2\frac{(1-t)^2}{f(Q(t))}\}\\ &= Q(t)\{H_F^{-1}(t) + 2\frac{(1-t)^2}{f(Q(t))}\} - 2H_F^{-1}(t)\frac{(1-t)^2}{f(Q(t))}\end{aligned} \tag{8.2}$$

for later purposes. Note that $Q(t) \geq H_F^{-1}(t)$ for any $t \in [0,1]$.

We were not able to identify $T_F(\cdot)$ as a transformed known process like, for example, the limit process of the multiplied mean residual life process $v_n(x)$ of (4.2) was identified by Hall and Wellner (1979) as a rescaled Wiener process. Thus we are unable to construct asymptotic confidence <u>bands</u> for H_F^{-1} on the basis of our Theorems 6.2 or 6.3, as they did for M_F. However, the resampling method of Section 17 gives an alternative solution. We shall shortly see in point 5) below that the scaled total time on test process is specifically tailored to fit the exponential distribution. Otherwise we are left with drawing pointwise inferences only from our results. This line is taken up in point 7) below.

4) The limit process of the scaled total time on test process $s_n(u)$, $0 \le u \le 1$, under the conditions of Theorem 7.2,

$$\mathbf{S}(u) = S_F(u) = \mu^{-1}T_F(u) - \mu^{-2}H_F^{-1}(u)T_F(1), \quad 0 \le u \le 1,$$

has covariance

$$\begin{aligned}\sigma_2(s,t) &= ES_F(s)S_F(t) \\ &= \mu^{-2}\sigma_1(s,t) + \mu^{-4}H_F^{-1}(s)H_F^{-1}(t)\sigma_1(1,1) \\ &\quad -\mu^{-3}H_F^{-1}(t)\sigma_1(s,1) - \mu^{-3}H_F^{-1}(s)\sigma_1(t,1).\end{aligned}$$

This appears even more complicated, at the first sight, than $T_F(\cdot)$. Note, however, that $H_{F_{\theta,\sigma}}^{-1}(u) = \sigma H_F^{-1}(u) + \theta$, $0 \le u \le 1$, and

$$\int_0^\infty x \, dF_{\theta,\sigma}(x) = \sigma\mu + \theta ,$$

and hence by (8.1),

$$S_{F_\sigma}(u) \equiv S_{F_{0,\sigma}}(u) \equiv S_F(u) \quad \text{for} \quad \theta = 0.$$

So, while $T_F(\cdot)$ is shift-free, $S_F(\cdot)$ is scale-free.

5) The limit process $S_F(\cdot)$ for the scaled total time on test process $s_n(u)$, $0 \le u \le 1$, reduces to a well-known process if the underlying distribution is exponential.

Indeed, assume that $F(x) = 1-\exp(-x/\mu)$, $x \ge 0$, $F(x)=0$, $x < 0$, with some positive mean value μ. Then $Q(y) = -\mu \log(1-y)$, $H_F^{-1}(y) = \mu y$, and hence $D_F^{-1}(y) \equiv y$, $0 \le y \le 1$. In fact, the converse is also true in the following sense. Consider the following class of distribution functions F_μ :

$$\mathcal{F} = \{F_\mu : F(0+) = 0 \quad \mu = \int_0^\infty x \, dF_\mu(x) < \infty\}.$$

If for $F_\mu \in \mathcal{F}$ we have $D_F^{-1}(y) \equiv y$ on $[0,1]$, then a simple differentiation argument yields that $F_\mu(x) = 1-\exp(-x/\mu)$ for $x \ge 0$. This characterisation result must be known.

Assuming now this exponentiality, our limiting process is

$$S_{F_\mu}(u) = \int_0^u \frac{B(y)}{1-y}\,dy + B(u) - u\int_0^1 \frac{B(y)}{1-y}\,dy. \tag{8.3}$$

This process has been identified in M. Csörgő and Révész (1981a, 1983) (see also Deheuvels (1982) and point 6) below) as a Brownian bridge on [0,1]. Introduce then the Kolmogorov

$$K_1(x)=\begin{cases}\sum_{k=-\infty}^{\infty}(-1)^k\exp(-k^2x^2), & x\geq 0,\\ 0, & x<0,\end{cases}$$

the Smirnov

$$K_2(x)=\begin{cases}1-e^{-2x^2}, & x\geq 0\\ 0, & x<0,\end{cases}$$

and the Cramér-von Mises-Smirnov

$$K_3(x)=\begin{cases}1-\frac{2}{\pi}\sum_{k=1}^{\infty}(-1)^{k+1}\int_{(2k-1)\pi}^{2k\pi}\frac{\exp(-u^2x/2)}{\sqrt{-u\sin u}}\,du, & x\geq 0,\\ 0, & x<0,\end{cases}$$

$$=\begin{cases}\frac{1}{\pi\sqrt{x}}\sum_{k=0}^{\infty}(-1)^k\binom{-1/2}{k}\sqrt{4k+1}\exp\left(-\frac{(4k+1)^2}{16x}\right)B_{1/4}\left(\frac{(4k+1)^2}{16x}\right), & x\geq 0,\\ 0, & x<0,\end{cases}$$

distribution functions for the corresponding functionals of the Brownian bridge:

$$K_1(x)=\mathrm{pr}\{\sup_{0\leq y\leq 1}|B(y)|\leq x\},$$

$$K_2(x)=\mathrm{pr}\{\sup_{0\leq y\leq 1}B(y)\leq x\},$$

$$K_3(x)=\mathrm{pr}\{\int_0^1 B^2(y)\,dy\leq x\},$$

where the second form of K_3 is due to Anderson and Darling (1952) and where $B_{1/4}$ is a standard Bessel function of parameter 1/4. All these functions are extensively tabulated in many textbooks. The most precise table for K_3 appears to be in Martynov (1978). We may also consider Kuiper's (1960) distribution function

$$K_4(x)=\mathrm{pr}\{\sup_{0\leq y\leq 1}B(y)-\inf_{0\leq y\leq 1}B(y)\leq x\}$$

$$=\begin{cases}1-2\sum_{k=1}^{\infty}(4k^2x^2-1)\exp(-2k^2x^2), & x\geq 0,\\ 0, & x<0.\end{cases}$$

For any exponential distribution F_μ, condition (6.13) is satisfied with $\alpha = 0$ and $\beta = 1$, moreover $J_1 = 1$ in condition (6.14). Hence Theorem 6.3 implies the following results.

COROLLARY 8.1. *If* $F(x) = 1 - \exp(-x/\mu)$, $x \geq 0$, *with some* $\mu > 0$, *then*

$$\sup_{-\infty<x<\infty} |\mathrm{pr}\{n^{\frac{1}{2}} \sup_{0\leq y\leq 1} |D_n^{-1}(y)-y| \leq x\} - K_1(x)| \longrightarrow 0,$$

$$\sup_{-\infty<x<\infty} |\mathrm{pr}\{n^{\frac{1}{2}} \sup_{0\leq y\leq 1} (D_n^{-1}(y)-y) \leq x| - K_2(x)| \longrightarrow 0,$$

$$\sup_{-\infty<x<\infty} |\mathrm{pr}\{n \int_0^1 (D_n^{-1}(y)-y)^2 dy \leq x\} - K_3(x)| \longrightarrow 0,$$

$$\sup_{-\infty<x<\infty} |\mathrm{pr}\{n^{\frac{1}{2}}[\sup_{0\leq y\leq 1} (D_n^{-1}(y)-y) - \inf_{0\leq y\leq 1} (D_n^{-1}(y)-y)] \leq x\} - K_4(x)| \longrightarrow 0.$$

These results can be used for testing the <u>composite</u> hypothesis of exponentiality with zero shift. If F is not exponential, then

$$n^{\frac{1}{2}}(D_n^{-1}(y)-y) = s_n(y) + n^{\frac{1}{2}}(D_F^{-1}(y)-y)$$

and hence, by the above characterisation result,

$$n^{\frac{1}{2}} \sup_{0\leq y\leq 1} |D_n^{-1}(y)-y| \xrightarrow{a.s.} \infty$$

and the same is true for the corresponding Smirnov and Kuiper statistics. Moreover,

$$n \int_0^1 (D_n^{-1}(y)-y)^2 dy \xrightarrow{a.s.} \infty$$

whenever

$$\int_0^1 (D_F^{-1}(y)-y)^2 dy > 0$$

for any F satisfying the conditions of Theorem 6.2. Thus the tests are consistent against the indicated alternatives. In fact the performance of some closely related statistics has already been studied extensively by M. Csörgő, Seshadri and Yalovsky (1975) against several alternatives. See also Doksum and Yandell (1984).

If the underlying distribution is Weibull, $F(x) = 1-\exp(-(x/\sigma)^m)$, $x \geq 0$, with a <u>known</u> shape parameter $m > 0$, then, of course, Corollary 8.1 remains true with $D_n^{-1}(y)$ replaced by $D_n^{-1}(y;m)$, where the latter scaled empirical total time on test function is constructed from the transformed data $X_1^m,\ldots,X_n^m$.

We conjecture that the limiting Gaussian process $\{S_F(u): 0 \leq u \leq 1\}$ is a Brownian bridge <u>only</u> in the case of exponential densities, i.e., the only solutions of the covariance integral equation

$$\sigma_1(s,t) = ES_F(s)S_F(t) = \min(s,t) - st$$

are the density functions $f_\mu(x) = \frac{1}{\mu}\exp(-x/\mu)$, $x \geq 0$.

6) Suppose, but only for the present point and Section 10 below, that the random variables $X_1, X_2, \ldots$ of our basic sequence are not necessarily nonnegative. Consider the scale and shift family of distribution functions

$$(8.4) \quad \mathcal{F} = \{F(\cdot;\theta,\sigma) : F(x;\theta,\sigma) \equiv F_o(\tfrac{x-\theta}{\sigma}), \quad -\infty < \theta < \infty, \ \sigma > 0\},$$

where F_o is a specified generic distribution function whose density function $f_o(x) \equiv F_o'(x)$ is positive and has a continuous derivative on the open support $(a_o, b_o) = (t_{F_o}, T_{F_o})$ of F_o, $-\infty \leq a_o < b_o \leq \infty$. Let us also assume that

$$A_o = \lim_{x \downarrow a_o} f_o(x), \quad B_o = \lim_{x \uparrow b_o} f(x) < \infty ,$$

and either $\min(A_o, B_o) < \infty$, or if $A_o = 0$ (resp. $B_o = 0$), then f_o is nondecreasing (resp. nonincreasing) on an interval to the right of a_o (resp. to the left of b_o). Finally, we assume that condition (6.4) is satisfied for F_o, that is,

$$\sup_{-\infty < x < \infty} F_o(x)(1-F_o(x)) \frac{|f_o'(x)|}{f_o^2(x)} < \infty .$$

We remark that these are the conditions on F_o under which the quantile process belonging to F_o was originally strongly approximated by Brownian bridges in M. Csörgő and Révész (1978) (cf. also M. Csörgő and Révész (1981)). Various improvements of this result are found in M. Csörgő, S. Csörgő, Horváth and Révész (1984).

Aiming at a test of the composite hypothesis $H : F \in \mathcal{F}$ and following upon a conjecture of Parzen (1979), M. Csörgő and Révész (1981a, 1983) considered the following processes:

$$p_n(u) = n^{\frac{1}{2}}\left\{\frac{\sum_{k=1}^{[nu]-1} f_o(Q_o(\frac{k}{n+1}))(X_{k+1:n} - X_{k:n})}{\sum_{k=1}^{n-1} f_o(Q_o(\frac{k}{n+1}))(X_{k+1:n} - X_{k:n})} - u\right\}$$

for $u \in [2/n, 1]$ and $p_n(u) = 0$ for $u \in [0, 2/n]$, where $Q_o = F_o^{-1}$. This process is clearly scale and shift-free for each n, and under the above conditions and H it was shown to converge weakly (relative to the supremum norm) to the zero-mean Gaussian process

$$\tilde{G}_{F_o}(u) = B(u) - \int_0^u \frac{f_o'(Q_o(y))}{f_o^2(Q_o(y))} B(y)dy + u\int_0^1 \frac{f_o'(Q_o(y))}{f_o^2(Q_o(y))} B(y)dy,$$

$0 \le u \le 1$, depending on the generic distribution F_o. This is clearly the Brownian bridge if F_o is the uniform distribution on $(0,1)$. If $F_o(x) = 1-\exp(-x)$, $x \ge 0$, then $\tilde{G}_{F_o}$ reduces to the right hand side of (8.3), and that process is again a Brownian bridge. (This is also the case if $F_o(x) = \exp(x)$, $x \le 0$, and Deheuvels (1982) has shown that these are the only three cases when $\tilde{G}_{F_o}$ reduces to a Brownian bridge.)

It is not surprising that $\{p_n(u),\ 0 \le u \le 1\}$ and the scaled total time on test processes are asymptotically equivalent if the underlying distribution is exponential with zero shift. Since $f_o(Q_o(u)) = 1-u$ if $f_o(x) = \exp(-x)$, $x \ge 0$, in the notation of the Introduction we may write

$$p_n(u) = n^{\frac{1}{2}} \left\{ \frac{\sum_{k=2}^{[nu]} W_{k:n}}{\sum_{k=2}^{n} W_{k:n}} - u \right\}, \quad \frac{2}{n} \le u \le 1,$$

and

$$s_n(u) = n^{\frac{1}{2}} \left\{ \frac{\sum_{k=1}^{[nu]+1} W_{k:n}}{\sum_{k=1}^{n} W_{k:n}} - u \right\}, \quad 0 \le u < 1$$

with $W_{k:n} = (n+1-k)(X_{k:n} - X_{k-1:n})$, $k = 1,\ldots,n$, $X_{0:n} \equiv 0$. Both processes are scale-free, and since $W_{1:n} = nX_{1:n}$ is not included in $p_n(u)$, it is also shift-free. This is not true for $s_n(u)$, but when the underlying distribution is exponential with any scale parameter and zero shift, then $\sup\{|s_n(u) - p_n(u)| : 0 \le u \le 1\} \to 0$ in probability.

The phenomenon occurring here clearly calls for a simple modification of the scaled total time on test process in general which will be scale and shift free. This modification is considered in the next section, following the last point here.

7) As noted in the last paragraph of point 3) above, we can base statistical inference on the obtained results only pointwise, apart from the exponential distribution. For another possible approach we refer again to Section 17.

Let us choose a fixed point $u \in (0,1)$ and consider the following estimator

$$\sigma_{1n}^2(u) = Q_n(u)H_n^{-1}(u) + 2Q_n(u)\phi_n(u)(1-u)^2$$
$$- 2H_n^{-1}(u)(1-u)^2\phi_n(u)$$

for $\sigma_1^2(u)$ of (8.2), where

$$\phi_n(u) = n^{\delta} \int_0^1 \lambda(n^{\delta}(y-u))\,dQ_n(y), \quad 0 < \delta < 1/2,$$

with a density function λ vanishing outside $(-1/2,1/2)$ and having zero expectation, a finite second moment and a bounded derivative, and where

$$(8.5) \qquad Q_n(y) = \begin{cases} X_{k:n}, & \frac{k-1}{n} \le y < \frac{k}{n}, \quad k=1,\dots,n, \\ X_{n:n}, & y = 1 \end{cases}$$

is the empirical quantile function of the sample.

THEOREM 8.2. (i) _If_ $\mu < \infty$, _and_ Q _is continuous at_ u _then_

$$H_n^{-1}(u) \xrightarrow{a.s.} H_F^{-1}(u).$$

(ii) _If_ $\mu < \infty$, $f(Q(\cdot))$ _is continuous at_ u _and_ $f(Q(u)) > 0$, _then_

$$\lim_{n\to\infty} \mathrm{pr}\left\{\frac{t_n(u)}{\sigma_1(u)} \le x\right\} = \Phi(x), \quad -\infty < x < \infty,$$

where Φ _is the standard normal distribution function._

(iii) _If_ $\mu < \infty$, $Q(\cdot)$ _is continuous on_ $[0, u+\varepsilon)$ _with any small_ $\varepsilon > 0$, f _is continuous in a neighbourhood of_ $Q(u)$ _and_ $f(Q(u)) > 0$, _then_

$$\lim_{n\to\infty} \mathrm{pr}\{H_n^{-1}(u) - x\frac{\sigma_{1n}(u)}{\sqrt{n}} < H_F^{-1}(u) < H_n^{-1}(u) + x\frac{\sigma_{1n}(u)}{\sqrt{n}}\} = 2\Phi(x)-1$$

for any x _on the line._

Proof. Part (i) follows trivially from the proof of Theorem 6.1. As we have already remarked, this statement was first proved by Langberg, León and Proschan (1980). As to Part (ii), a simple inspection of the proof of Theorem 6.2 shows that $|t_n(u)-T_n(u)| \to 0$ in probability on our basic space (Ω,A,P) only under the stated conditions, and this implies the statement.

Finally, turning now to the proof of Part (iii), we first note that if Q is continuous at $u \in (0,1)$ then $Q_n(u) \to Q(u)$ almost surely. A glance at the proof of Theorem 6.1 will show that under the conditions

$$(8.6) \qquad \mu < \infty \text{ and } Q(\cdot) \text{ is continuous on } [0,u+\varepsilon),$$

$H_n^{-1}(u) \to H^{-1}(u)$ almost surely. The proof of Theorem 2 of M. Csörgő and Révész (1980) implies that $\phi_n(u) \to 1/f(Q(u))$ almost surely under the conditions of Part (ii) only. Consequently we have $\sigma_{1n}^2(u) \to \sigma^2(u)$ almost surely, and hence also the asymptotic confidence interval statement of Part (iii).

9. TOTAL TIME ON TEST FROM THE FIRST FAILURE.

9.1. Right-sided distributions. Let us allow for the possibility that the lower end of the support of F,

$$t_F = \sup\{t : F(t) = 0\},$$

is not necessarily zero, but

$$-\infty < t_F . \tag{9.1}$$

We define the nth total time on test function from the first failure on as

$$N_n^1(u) = \begin{cases} 0 & , \quad u < 1/n, \\ \frac{1}{n} \sum_{k=2}^{[nu]+1} W_{k:n} & , \quad 1/n \le u < 1, \\ \frac{1}{n} \sum_{k=2}^{n} W_{k:n} & , \quad u = 1. \end{cases}$$

With t_F as in (9.1), the corresponding theoretical function (cf. (1.5)) is

$$N_F^1(u) = H_F^{-1}(u) - t_F = \int_0^u (1-y)\, dQ(y), \quad 0 \le u \le 1,$$

and let

$$\eta_n^1(u) = n^{\frac{1}{2}}\{N_n^1(u) - N_F^1(u)\}, \quad 0 \le u \le 1,$$

be the corresponding normalized process. With $T_n(\cdot)$ denoting the same sequence of copies of $T_F(\cdot)$ as in Section 6 above, the following result is rather immediate.

THEOREM 9.1. (i) *If the conditions of Theorem 6.1 are satisfied, then*

$$\Delta_n^{(7)} = \sup_{0 \le u \le 1} |N_n^1(u) - N_F^1(u)| \xrightarrow{a.s.} 0.$$

(ii) *If the conditions of Theorem 6.2 are satisfied, then*

$$\Delta_n^{(8)} = \sup_{0 \le u \le 1} |\eta_n^1(u) - T_n(u)| \xrightarrow{P} 0 .$$

(iii) *If the conditions of Theorem 6.3 are satisfied, then*

$$\Delta_n^{(8)} = \sup_{0 \le u \le 1} |\eta_n^1(u) - T_n(u)| \overset{a.s.}{=} O(n^{-\tau})$$

with the same τ *as in Theorem 6.3.*

Proof. (i) We have

$$\Delta_n^{(7)} \leq \sup_{0\leq u\leq 1} |H_n(u) - H_F^{-1}(u)| + |X_{1:n} - t_F|$$

$$= \Delta_n^{(5)} + (Q(U_{1:n}) - Q(0))$$

and the result follows by Theorem 6.1 and the continuity of Q.

(ii) We have, with the usual $\hat{h}(t) = \inf\{h(s):0<s\leq t\}$ (cf. (2.24) for h),

$$\Delta_n^{(8)} \leq \Delta_n^{(6)} + n^{\frac{1}{2}}(Q(U_{1:n}) - Q(0))$$

$$\leq \Delta_n^{(6)} + 2C_{11}(\hat{h}(U_{1:n}))^{-1}(nU_{1:n})^{\frac{1}{2}}$$

for large enough n, as in the proof of Theorem 6.2, and case (ii) follows as in the proof of Theorem 6.2.

(iii) Similarly, and by (6.17),

$$\Delta_n^{(8)} \leq \Delta_n^{(6)} + (1-\alpha)^{-1}C_{13}(nU_{1:n})^{1-\alpha}n^{\alpha-\frac{1}{2}},$$

and case (iii) follows as in the lines between (6.17) and (6.18) of the proof of Theorem 6.3.

Consider now the scaled version of the above process:

$$\xi_n^1(u) = n^{\frac{1}{2}}\{J_n^1(u) - J_F^1(u)\},\quad 0 \leq u \leq 1,$$

with

$$J_n^1(u) = \frac{N_n^1(u)}{N_n^1(1)},\quad J_F^1(u) = \frac{N_F^1(u)}{N_F^1(1)},\quad 0 \leq u \leq 1.$$

Since the role of $\mu = N_F^1(1)+t_F$ is now taken over by

$$N_F^1(1) = \int_0^1 (1-y)\,dQ(y),$$

we consider the sequence $\{\Xi_n^1(\cdot)\}_{n=1}^{\infty}$ of copies of the zero-mean Gaussian process

$$\Xi_F^1(u) = \frac{1}{N_F^1(1)} T_F(u) - \frac{N_F^1(u)}{(N_F^1(1))^2} T_F(1),\quad 0 \leq u \leq 1.$$

Exactly as Theorems 6.1,2,3 implied Theorems 7.1,2,3, Theorem 9.1 now implies

THEOREM 9.2. (i) Under the conditions of Theorem 6.1

$$\sup_{0\leq u\leq 1} |J_n^1(u) - J_F^1(u)| \xrightarrow{a.s.} 0.$$

(ii) <u>Under the conditions of Theorem 6.2</u>

$$\sup_{0\leq u\leq 1} |\xi_n^1(u) - \Xi_n^1(u)| \xrightarrow{P} 0.$$

(iii) <u>Under the conditions of Theorem 6.3</u>

$$\sup_{0\leq u\leq 1} |\xi_n^1(u) - \Xi_n^1(u)| \overset{a.s.}{=} O(n^{-\tau})$$

<u>with the same</u> τ <u>as in Theorem 6.3</u>.

The reason for introducing the above scaled total time on test process $\xi_n^1(\cdot)$ from the first failure on is obviously the fact that $J_n^1(\cdot)$ above is invariant under scale and shift transformation of the data. Consider now the family F in (8.4) with a generic right-sided distribution function F_o, i.e., now

$$t_{F_o} > -\infty \text{ is assumed.}$$

Then $N_F^1(\cdot)$ is the same as $N_{F_o}^1(\cdot)$ for any $F \in F$, and so the whole sequence of processes $\{\xi_n^1(\cdot)\}$ is the same for any $F \in F$, and $\Xi_F^1(\cdot)$ is the same for any $F \in F$.

The covariance function of $\Xi_F^1(\cdot)$ is obtained from that of $S_F(\cdot)$ by replacing $H_F^{-1}(\cdot)$ by $N_F^1(\cdot)$ and μ by $N_F^1(1)$. This function is again hopelessly complicated for any F_o other than $F_o(x) = 1-\exp(-x)$, $x \geq 0$. In the latter case $\Xi_{F_o}^1$ is a Brownian bridge, and hence we have the following consequence of Theorem 9.2(iii).

COROLLARY 9.3. <u>If</u> $F(x) = 1-\exp(-(x-\theta)/\sigma)$ <u>with some real</u> θ <u>and</u> $\sigma > 0$, <u>then the four statements of Corollary 8.1 hold true with</u> $n^{\frac{1}{2}}(D_n^{-1}(y)-y)$ <u>replaced by</u> $n^{\frac{1}{2}}(J_n^1(y)-y)$.

9.2. <u>General two-sided distributions</u>. In the present subsections we allow that even $t_F = -\infty$ may occur. The first question is then: what is the two-sided analogue of the total time on test transform of F, or, rather, what is the two-sided analogue of N_F^1 above? A possible answer to this question is the following definition:

$$N_F(u) = \begin{cases} \int_0^u y\,dQ(y) & , \text{ if } 0 \leq u < t_o, \\ \int_0^{t_o} y\,dQ(y) + \int_{t_o}^u (1-y)\,dQ(y), & \text{if } t_o \leq u \leq 1, \end{cases}$$

where t_o is some fixed value in $[0,1]$ such that $-\infty < Q(t_o) < \infty$.

If $t_F > -\infty$, then the choice $t_o = 0$ leads to the above N_F^1, while if $T_F = \inf\{t : F(t)=1\} < \infty$ then the choice $t_o = 1$ gives the right-sided analogue

$$N_F^2(u) = \int_0^u y\,dQ(y), \quad 0 \le u \le 1,$$

of N_F^1.

The empirical counterpart of $N_F(\cdot)$ is

$$N_n(u) = \begin{cases} \frac{1}{n} \sum_{k=1}^{[nu]+1} W_{k:n}^*, & 0 \le u \le 1, \\ \frac{1}{n} \sum_{k=1}^{n} W_{k:n}^*, & u = 1, \end{cases}$$

where

$$W_{k:n}^* = \begin{cases} k(X_{k+1:n} - X_{k:n}) & , 1 \le k \le k_o, \\ W_{k:n} = (n+1-k)(X_{k:n} - X_{k-1:n}), & k_o < k \le n, \end{cases}$$

and where $k_o \in [1,n]$ is that integer for which

$$\frac{k_o - 1}{n} < t_o \le \frac{k_o}{n}.$$

The same type of elementary computations that led to (6.1) now show that $N_n(\cdot)$ can be represented by the integral

$$\begin{cases} \int_0^{U_n(u)} E_n(y)\,dQ(y) & , 0 \le u \le t_o, \\ \int_0^{U_n(t_o)} E_n(y)\,dQ(y) + \int_{U_n(t_o)}^{U_n(u)} (1-E_n(y)\,dQ(y), & t_o \le u \le 1. \end{cases}$$

If $t_F > -\infty$, then the choice $t_o = 0$ makes $N_n(\cdot)$ to become $N_n^1(\cdot)$ above. If $T_F < \infty$, then the choice $t_o = 1$ reduces $N_n(\cdot)$ to $N_n^2(\cdot)$, which is the empirical counterpart of N_F^2 of (9.1) above. Of course, $N_n^2(\cdot)$ is exactly the $N_n^1(\cdot)$ corresponding to $-X_1,\ldots,-X_n$. Hence all the results so far, proved for $n^{\frac{1}{2}}(N_n^1(\cdot)-N_F^1(\cdot))$ or for its scaled version, have the easily formulated left-sided analogues for $\eta_n^2(\cdot) = n^{\frac{1}{2}}(N_n^2(\cdot)-N_F^2(\cdot))$ and $\xi_n^2(\cdot) = n^{\frac{1}{2}}(J_n^2(\cdot)-J_F^2(\cdot))$, the latter being the scaled version of the former. Specifically, $\xi_n^2(\cdot)$ converges weakly in the space of continuous functions on $[0,1]$ to the Brownian bridge if the underlying distribution function is $F(x) = \exp((x-\theta)/\sigma)$, $x \le \theta$, with some θ and $\sigma > 0$.

Because of the preceding paragraph, we shall effectively assume that $t'_F = -\infty$ and $T_F = \infty$. This assumption enables us to state our two-sided results in their sharpest form.

Introducing the scaled two-sided "total time on test process from the first failure on" as

$$\xi_n(u) = n^{\frac{1}{2}}\{J_n(u) - J_F(u)\}, \quad 0 \le u \le 1,$$

with

$$J_n(u) = \frac{N_n(u)}{N_n(1)}, \quad J_F(u) = \frac{N_F(u)}{N_F(1)},$$

depending also on t_o, consider the following sequence of zero-mean Gaussian processes:

$$\Xi_n(u) = \frac{1}{N_F(1)} \{\Theta_n(u) - \frac{N_F(u)}{N_F(1)} \Theta_n(1)\}, \quad 0 \le u \le 1,$$

where Θ_n is a sequence of copies of the Gaussian processes

$$\Theta_F(u) = -\int_0^u B(y)dQ(y) - \frac{u}{f(Q(u))} B(u), \quad \text{if } 0 \le u \le t_o,$$

and

$$\Theta_F(u) = -\int_0^{t_o} B(y)dQ(y) - \frac{t_o}{f(Q(t_o))} B(t_o)$$

$$+ \int_{t_o}^u B(y)dQ(y) - \frac{1-t_o}{f(Q(t_o))} B(t_o) + \frac{1-u}{f(Q(u))} B(u), \quad \text{if } t_o \le u \le 1,$$

obtained by replacing B with B_n.

When approximating $\eta_n(u) = n^{\frac{1}{2}}\{N_n(u) - N_F(u)\}$ by $\Theta_n(u)$, we cut the processes into two pieces at t_o. Results above t_o follow as in the proof of Theorem 9.1, and everything works also symmetrically below t_o if we assume the corresponding conditions to hold (symmetrically) around zero. Following this, the results for the scaled version $\xi_n(\cdot)$ are obtained in the usual way.

THEOREM 9.4. (i) *If* $\mu = EX < \infty$ *and* $Q = F^{-1}$ *is continuous on* $(0,1)$, *then*

$$\sup_{0<u<1} |J_n(u) - J_F(u)| \xrightarrow{a.s.} 0.$$

(ii) *Suppose that the density function* $f = F'$ *is continuous and positive on the open support of* F *and* $EX^2 < \infty$. *If, moreover,*

$$\sup_{0 \le u \le 1} \frac{uq(u)(1-u)}{f(Q(u))} < \infty$$

for some O'Reilly weight function q, *then*

$$\sup_{0\le u\le 1} |\xi_n(u) - \Xi_n(u)| \xrightarrow{P} 0.$$

(iii) <u>Suppose that the density function</u> $f = F'$ <u>is positive on the open support of</u> F <u>and that condition</u> (6.14) <u>holds for the derivative</u> $f' = F''$ <u>of the density function. If, moreover, condition</u> (6.13) <u>also holds with</u> $0 \le \alpha, \beta \le 1 + (1/8^{\frac{1}{2}})$, <u>then</u>

$$\sup_{0<u<1} |\xi_n(u) - \Xi_n(u)| \overset{a.s.}{=} O(n^{-\tau})$$

<u>for any</u> $\tau < \min(\tau_\alpha, \tau_\beta)$, <u>where</u>

$$\tau_\gamma = \begin{cases} \min(\frac{1}{4}, \frac{3-2\gamma}{10-4\gamma}), & \text{if } 0 \le \gamma \le 1, \\ \min(\frac{1}{4}, \frac{3-2\gamma}{10-4\gamma}, \frac{1-8(\gamma-1)^2}{8(\gamma-1)^2+4}), & \text{if } 1 < \gamma \le 1+8^{-\frac{1}{2}}, \end{cases}$$

<u>for</u> $\gamma = \alpha, \beta$.

Suppose now that the general scale and shift family of (8.4) is such that $t_{F_o} = -\infty$, $T_{F_o} = \infty$, and F_o satisfies the conditions of Theorem 9.4(ii) above. Then, if $F \in \mathcal{F}$, the process $\xi_n(\cdot)$ converges weakly to $\Xi_{F_o}(\cdot)$ with respect to the supremum topology, where $\Xi_{F_o}(\cdot)$ is obtained upon replacing $\Theta_n(\cdot)$ by $\Theta_{F_o}(\cdot)$ in the above definition of $\Xi_n(\cdot)$. Hence the random variables

$$(9.2) \qquad \sup_{0\le y\le 1} \xi_n(y), \quad \sup_{0\le u\le 1} |\xi_n(y)|, \quad \int_0^1 \xi_n^2(y)\,dy$$

converge in distribution to

$$(9.3) \qquad \sup_{0\le y\le 1} \Xi_{F_o}(y), \quad \sup_{0\le y\le 1} |\Xi_{F_o}(y)|, \quad \int_0^1 (\Xi_{F_o}(y))^2 dy$$

respectively. It follows easily from Part (ii) of Theorem 9.4 that the tests based on any of the three statistics in (9.2) are consistent against any alternative $F \notin \mathcal{F}$ that satisfies the conditions of Theorem 9.4(ii). The main difficulty is of course that we do not know the distributions of the random variables in (9.3) for any specified F_o. For further considerations we refer to Section 17.

Let F_o be the double exponential distribution function

$$F_o(x) = \begin{cases} \frac{1}{2} e^x, & x \le 0, \\ 1 - \frac{1}{2} e^{-x}, & x > 0. \end{cases}$$

In view of its symmetry, the obvious choice for t_o here is $1/2$, where $Q_o(1/2) = 0$. Now by simple computation $J_F(u) \equiv u$ on $[0,1]$

for any $F = F_{\theta,\sigma} \in \mathcal{F}$ corresponding to this double exponential F_o, since $N_F(u) = \sigma u/2$. The limiting process is

$$\Xi_{F_o}(u) = -2 \int_0^u \frac{B(y)}{y} dy - 2B(u)$$

$$+ u\{\int_0^{1/2} \frac{B(y)}{y} dy - \int_{1/2}^1 \frac{B(y)}{1-y} dy + 2B(\tfrac{1}{2})\}$$

for $0 \leq u \leq 1/2$, and

$$\Xi_{F_o} = -2 \int_0^{1/2} \frac{B(y)}{y} dy - 4B(\tfrac{1}{2})$$

$$+ 2 \int_{1/2}^u \frac{B(y)}{1-y} dy + 2B(u)$$

$$+ 2u\{\int_0^{1/2} \frac{B(y)}{y} dy - \int_{1/2}^1 \frac{B(y)}{1-y} dy + 2B(\tfrac{1}{2})\}$$

for $1/2 \leq u \leq 1$. We conjecture that this Ξ_{F_o} is a Brownian bridge on $[0,1]$.

9.3. <u>An estimate for the scale parameter in a scale and shift family</u>. Although the content of the present subsection could be formulated for general two-sided distributions on the basis of the preceding subsection, for the sake of simplicity we return to the right-sided setup of subsection 9.1. So consider again

$$\mathcal{F} = \{F(\cdot) = F_{\theta,\sigma}(\cdot) : F(x) \equiv F_o(\tfrac{x-\theta}{\sigma}), \quad -\infty < \theta < 0, \ \sigma > 0\},$$

with

$$t_{F_o} > -\infty .$$

With $Q_o = F_o^{-1}$, consider the scale estimator

$$\sigma_n = \frac{\sum_{k=2}^n W_{k:n}}{\int_0^1 (1-y) dQ_o(y)} = \frac{\frac{1}{n} \sum_{i=1}^n X_i - X_{1:n}}{\mu_o - t_{F_o}},$$

where

$$\mu_o = \int_{-\infty}^{\infty} x dF_o(x) .$$

A little arrangement in Theorem 9.1 shows that

$$\sigma_n = \sigma + n^{-\frac{1}{2}} \frac{T_n(1)}{\mu_o - t_{F_o}} + r_n,$$

where

$$T_n(1) = \int_0^1 B(y)\,dQ(y),$$

and

$$r_n = \begin{cases} o_P(n^{-\frac{1}{2}}), & \text{under the conditions of Part (ii),} \\ O(n^{-\frac{1}{2}-\tau}) \text{ a.s.,} & \text{under the conditions of Part (iii).} \end{cases}$$

Hence the following result is implied by Theorem 9.1, where X_o denotes a random variable with distribution function F_o.

COROLLARY 9.5. (i). _If_ $\mu_o < \infty$ _and_ Q _is continuous, then_ $\sigma_n \to \sigma$ _almost surely_.

(ii). _If the conditions of Theorem 6.2 are satisfied, then the distribution of_ $n^{\frac{1}{2}}(\sigma_n - \sigma)$ _converges to the normal_

$$N\left(0,\ \sigma^2 \frac{\operatorname{var}(X_o)}{(\mu_o - t_{F_o})^2}\right)$$

distribution.

(iii). _If the conditions of Theorem 6.3 are satisfied, then_

$$\sigma_n - \sigma \overset{a.s.}{=} O\left(\left(\frac{\log\log n}{n}\right)^{\frac{1}{2}}\right).$$

Of course when $F_o(x) = 1 - \exp(-x)$, $x \geq 0$, then the limiting variance of σ_n is σ^2.

On the other hand, since the second term in the limit process $T_F(\cdot)$ vanishes at $u = 1$, it is conceivable that the conditions of (ii) and (iii) can be relaxed. Indeed, since

$$\mu - t_F = \int_0^1 (1-y)\,dQ(y),$$

we have

$$n^{\frac{1}{2}}(\sigma_n - \sigma) = \frac{1}{\mu_o - t_{F_o}} \left\{\frac{1}{n^{\frac{1}{2}}} \sum_{k=1}^{n} (X_k - \mu) + n^{\frac{1}{2}}(t_F - X_{1:n})\right\},$$

where

$$n^{\frac{1}{2}}(t_F - X_{1:n}) = n^{\frac{1}{2}}(Q(0) - Q(U_{1:n})).$$

Hence, besides the existence of the variance, we have to assume our conditions on Q only around zero. So for the conclusion of (ii) in the above Corollary, condition (6.2) of Theorem 6.2 can be relaxed to

$$\limsup_{u \to 0} \frac{q(u)}{f(Q(u))} < \infty ,$$

and for the conclusion of (iii), the conditions (6.13) and (6.14) can be weakened to read as

$$\limsup_{u \to 0} \frac{u^{\alpha}}{f(Q(u))} < \infty \quad \text{with} \quad 0 \leq \alpha < 8^{-\frac{1}{2}}$$

and

$$\limsup_{u \to 0} \frac{u|f'(Q(u))|}{f^2(Q(u))} < \infty$$

respectively.

We note that the related scale-parameter estimator

$$\sum_{k=1}^{n-1} f(Q(\tfrac{k}{n+1}))(X_{k+1:n} - X_{k:n})$$

(cf. Point 6) in the preceding Section 8) was considered by M. Csörgő and Révész (1981a, 1983). Versions of the latter estimate date back to Weiss (1961, 1963).

10. UNSCALED EMPIRICAL LORENZ PROCESSES.

Let us recall that we assume the continuity of F, and consider the integral

$$\int_0^{U_n(y)} Q(x-)\,dE_n(x).$$

If $(k-1)/n \le y < k/n$ for some $k=1,\ldots,n$, then, apart from the set where either ties occur among $X_1,\ldots,X_n$ or they fall into discontinuity points of $Q(\cdot)$, we have

$$\begin{aligned}\int_0^{U_n(y)} Q(x-)\,dE_n(x) &= \int_0^{F(X_{k:n})} Q(x-)\,dE_n(x)\\ &= \frac{1}{n}\sum_{i=1}^{k} Q(U_{i:n}-)\\ &= \frac{1}{n}\sum_{i=1}^{[ny]+1} X_{i:n},\end{aligned}$$

and the same argument shows that the value of our integral at $y=1$ is $\bar{X}_n$ almost surely. Thus we arrive at our basic observation for the present section:

$$P\{\sup_{0\le y\le 1} |G_n(y) - \int_0^{U_n(y)} Q(x-)\,dE_n(x)| = 0\} = 1 \tag{10.1}$$

for each n, where

$$G_n(y) = \bar{X}_n L_n(y) = \begin{cases} \frac{1}{n}\sum_{i=1}^{[ny]+1} X_{i:n}, & 0 \le y < 1,\\ \bar{X}_n, & y = 1\end{cases}$$

is the unscaled empirical Lorenz curve. Introducing the unscaled Lorenz curve

$$G_F(y) = \mu L_F(y), \quad 0 \le y \le 1,$$

where L_n and L_F are as in (1.10) and (1.8), respectively, we have the following consistency result.

THEOREM 10.1. *If* $\mu = \int_0^1 Q(y)\,dy < \infty$, *then*

$$\Delta_n^{(7)} = \sup_{0<y<1} |G_n(y) - G_F(y)| \xrightarrow{a.s.} 0.$$

Proof. Let $\varepsilon > 0$ be arbitrarily small and choose $\beta \in (0,1)$ so large that

$$(10.2)\qquad I_1(\beta) = \int_{1-\beta}^{1} Q(x)dx < \varepsilon/2 .$$

The latter is possible by the proof of Lemma 1 on p. 148 of Feller (1966). Now by (10.1)

$$\Delta_n^{(7)} \overset{a.s.}{=} \sup_{0<y<1} \left|\int_0^{U_n(y)} Q(x-)dE_n(x) - \int_0^{y} Q(x-)dx\right|$$

$$\leq \sup_{0<y<1} \left|\int_0^{U_n(y)} Q(x-)dE_n(x) - \int_0^{U_n(y)} Q(x-)dx\right|$$

$$+ \sup_{0<y<1} |G_F(U_n(y)) - G_F(y)|,$$

where the second term goes to zero almost surely by the continuity of G_F and the first term is not greater than

$$\sup_{0<y<1-\beta} \left|\int_0^{y} Q(x-)dE_n(x) - \int_0^{y} Q(x-)dx\right| + \sup_{1-\beta\leq y<1} \left|\int_0^{y} Q(x-)dE_n(x) - \int_0^{y} Q(x-)dx\right|$$

$$\leq 2 \sup_{0<y\leq 1-\beta} \left|\int_0^{y} Q(x-)d(1-E_n(x)) - \int_0^{y} Q(x-)d(1-x)\right| + I_1(\beta) + \int_{1-\beta}^{1} Q(x)dE_n(x)$$

$$\leq 2\Delta_n^{(1)} + 2Q(1-\beta) \sup_{0<y<1} |E_n(y) - y|$$

$$+ I_1(\beta) + \frac{1}{n}\sum_{i=1}^{n} Q(U_i)\chi(\{U_i > 1-\beta\}),$$

upon integrating by parts in the last step, where $\Delta_n^{(1)}$ is that of Lemma 3.1. By this lemma, by Glivenko-Cantelli and by the strong law of large numbers we have $\limsup_{n\to\infty} \Delta_n^{(7)} \leq 2I_1(\beta) \leq \varepsilon$ a.s. for all $\varepsilon > 0$ and hence the theorem is proved.

THEOREM 10.2. <u>If</u> $Q = F^{-1}$ <u>is continuous on</u> $[0,1)$ <u>and</u> $EX^2 < \infty$, <u>then</u>

$$\Delta_n^{(8)} = \sup_{0\leq u\leq 1} |g_n(u) - \Gamma_n(u)| \xrightarrow{P} 0$$

<u>where</u>

$$g_n(u) = n^{1/2}(G_n(u) - G_F(u)), \quad 0 \leq u \leq 1,$$

<u>and the sequence</u> Γ_n <u>of zero-mean Gaussian processes is defined as</u>

$$(10.3)\qquad \Gamma_n(u) = \int_0^{u} B_n(y)dQ(y)$$

$$= Q(u)B_n(u) - \int_0^{u} Q(y)dB_n(y),$$

$0 \leq u \leq 1$, <u>with</u> B_n <u>as in</u> (2.6).

<u>Proof</u>. We have, almost surely,

$$
\begin{aligned}
\Delta_n^{(8)} &\leq \sup_{0\leq y\leq 1} \left|-\int_0^{U_n(y)} Q(x)\,d\alpha_n(x) + \int_0^{U_n(y)} Q(x)\,dB_n(x)\right| \\
&\quad + \sup_{0\leq y\leq 1} \left|\int_0^{y} Q(x)\,dB_n(x) - \int_0^{U_n(y)} Q(x)\,dB_n(x)\right| \qquad (10.4)\\
&\quad + \sup_{0\leq y\leq 1} \left|n^{1/2}\left\{\int_0^{U_n(y)} Q(x)\,dx - \int_0^{y} Q(x)\,dx\right\} - Q(y)B_n(y)\right| \\
&= \Delta_{1n}^* + \Delta_{2n}^* + \Delta_{3n}^* .
\end{aligned}
$$

Clearly,

$$
\begin{aligned}
\Delta_{1n}^* &\leq \sup_{0\leq y\leq 1} \left|\int_0^{y} Q(x)\,d(\alpha_n(x)-B_n(x))\right| \\
&\leq \sup_{0\leq y\leq 1} \left|\int_0^{y} (\alpha_n(x)-B_n(x))\,dQ(x)\right| \qquad (10.5)\\
&\quad + \sup_{0\leq y\leq 1} |Q(y)(\alpha_n(y)-B_n(y))| \\
&= \Delta_n^{(2)} + \Delta_{1n} ,
\end{aligned}
$$

and the condition that $EX^2 < \infty$ implies that $\Delta_n^{(2)} \overset{P}{\to} 0$ by Lemma 3.2, and Lemma 2.4 implies $\Delta_{1n} \overset{P}{\to} 0$.

Integrating by parts again,

$$
\Delta_{2n}^* \leq \Delta_n^{(4)} + \Delta_{2n} \qquad (10.6)
$$

where $\Delta_n^{(4)} \overset{P}{\to} 0$ by Lemma 5.2, and

$$
\Delta_{2n} = \sup_{0\leq y\leq 1} |Q(y)B_n(y)-Q(U_n(y))B_n(U_n(y))|. \qquad (10.7)
$$

Given $\varepsilon \in (0,1)$ we have

$$
\begin{aligned}
\Delta_{2n} &\leq \sup_{0\leq y\leq 1-\varepsilon} |Q(y)-Q(U_n(y))|\,|B_n(y)| \\
&\quad + \sup_{0\leq y\leq 1-\varepsilon} |B_n(y)-B_n(U_n(y))|\,|Q(U_n(y))| \\
&\quad + \sup_{1-\varepsilon\leq y\leq 1} |Q(y)B_n(y)| + \sup_{1-\varepsilon\leq y\leq 1} |Q(U_n(y))B_n(U_n(y))|.
\end{aligned}
$$

Here the first two terms go to zero in probability by the continuity of Q and the Brownian bridge, respectively, while the third term goes to zero in probability as $\varepsilon \to 0$ either by an application of Lemma 2.3, or directly by Lemma 2.5 and (2.18). The fourth term is less than or equal to

$$R_n^* = \sup_{U_n(1-\varepsilon)\le y\le 1} |Q(y)B_n(y)|,$$

where $U_n(1-\varepsilon) \to 1-\varepsilon$ a.s., and hence for any $\delta > 0$

$$\lim_{\varepsilon\to 0} \limsup_{n\to\infty} P\{R_n^* > \delta\} = 0.$$

Next we consider Δ_{3n}^* in (10.4). Let $0 < \varepsilon < 1$ be arbitrary and we assume that n is so large that $1/n < \varepsilon$. We have

$$\begin{aligned}
\Delta_{3n}^* \le & \sup_{0\le y\le 1-\varepsilon} \left|n^{\frac{1}{2}} \int_y^{U_n(y)} Q(x)\,dx - Q(y)u_n(y)\right| \\
& + \sup_{0\le y\le 1-\varepsilon} |Q(y)(u_n(y)-B_n(y))| \\
& + \sup_{1-\varepsilon\le y<1} \left|n^{\frac{1}{2}} \int_y^{U_n(y)} Q(x)\,dx\right| \qquad (10.8) \\
& + \sup_{1-\varepsilon\le y<1} |Q(y)B_n(y)| \\
= & \Delta_{3n} + \ldots + \Delta_{6n},
\end{aligned}$$

where

$$\begin{aligned}
\Delta_{3n} = & \sup_{0\le y\le 1-\varepsilon} \left|n^{\frac{1}{2}} \int_y^{U_n(y)} (Q(x)-Q(y))\,dx\right| \\
\le & \sup_{0\le y\le 1-\varepsilon} \; \sup_{y\wedge U_n(y)\le x\le y\vee U_n(y)} |Q(y)-Q(x)|\,|u_n(y)| \\
\le & \sup_{0\le y\le 1-\varepsilon} |Q(y)-Q(U_n(y))| \sup_{0\le y\le 1} |u_n(y)|.
\end{aligned}$$

This upper bound goes to zero in probability, for the second sup has a limiting distribution while the first one tends to zero in probability on account of the fact that

$$\sup_{0\le y\le 1} |y-U_n(y)| \xrightarrow{P} 0$$

and the uniform continuity of Q on $[0,1-\varepsilon]$. The next term is trivial:

$$\Delta_{4n} \leq Q(1-\varepsilon) \sup_{0\leq y\leq 1} |u_n(y)-B_n(y)| \xrightarrow{a.s.} 0.$$

Next,

$$\Delta_{5n} \leq \sup_{1-\varepsilon\leq y\leq 1-\frac{1}{n}} |n^{\frac{1}{2}} \int_y^{U_n(y)} Q(x)\,dx|$$

$$+ \sup_{1-\frac{1}{n}\leq y<1} |n^{\frac{1}{2}} \int_y^{U_n(y)} Q(x)\,dx|$$

$$\leq \sup_{1-\varepsilon\leq y\leq 1-\frac{1}{n}} \sup_{y\wedge U_n(y)\leq x\leq y\vee U_n(y)} Q(x)|u_n(y)|$$

$$+ n^{\frac{1}{2}} \int_{U_{n:n}}^{1} Q(x)\,dx + n^{\frac{1}{2}} \int_{1-\frac{1}{n}}^{1} Q(x)\,dx$$

$$\leq \sup_{1-\varepsilon\leq y\leq 1-\frac{1}{n}} Q(y)|u_n(y)| + \sup_{1-\varepsilon\leq y\leq 1-\frac{1}{n}} Q(U_n(y))|u_n(y)|$$

$$+ (n(1-U_{n:n}))^{\frac{1}{2}}\left(\int_{U_{n:n}}^{1} Q^2(x)\,dx\right)^{\frac{1}{2}} + \int_{1-\frac{1}{n}}^{1} Q^2(x)\,dx,$$

where, in the last step, we applied the Bunjakovskii-Schwarz inequality. Here, for the first term,

$$\lim_{\varepsilon\to 0} \limsup_{n\to\infty} P\{\sup_{1-\varepsilon\leq y\leq 1-\frac{1}{n}} Q(y)|u_n(y)| > \eta\}$$

$$= \lim_{\varepsilon\to 0} P\{\sup_{1-\varepsilon\leq y<1} Q(y)|B(y)| > \eta\} = 0$$

for any $\eta > 0$ by Lemmas 2.4 and 2.5 and (2.18). On the other hand, Lemma 2.7 implies that for each $\lambda > 1$,

$$P\{\frac{1-U_n(y)}{1-y} < \lambda,\quad 0 \leq y \leq 1-\frac{1}{n}\} \geq 1-\frac{1}{\lambda}$$

and therefore, for the second term in the latter upper bound for Δ_{5n},

$$\lim_{\varepsilon\to 0} \limsup_{n\to\infty} P\{\sup_{1-\varepsilon\leq y\leq 1-\frac{1}{n}} Q(U_n(y))|u_n(y)| > \eta\}$$

$$\leq \frac{1}{\lambda} + \lim_{\varepsilon\to 0} \limsup_{n\to\infty} P\{\sup_{1-\varepsilon\leq y\leq 1-\frac{1}{n}} Q(1+\tfrac{1}{\lambda}(y-1))|u_n(y)| > \eta\}$$

$$= \frac{1}{\lambda} + \lim_{\varepsilon\to 0} P\{\sup_{1-\varepsilon\leq y<1} Q(1+\tfrac{1}{\lambda}(y-1))|B(y)| > \eta\}$$

$$= \frac{1}{\lambda}.$$

This follows from the fact that $Q(1 + \frac{1}{\lambda}(y-1))$ is square-integrable, and hence an O'Reilly weight function on $[\frac{1}{2}, 1)$, for any $\lambda > 1$. Since λ is arbitrarily large, the said second term in question is arbitrarily small. The first factor random variable in the third term above for Δ_{5n} has a limiting distribution, while the second factor there goes to zero in probability since Q^2 is integrable. The fourth term of the last inequality for Δ_{5n} converges to zero by the same reason.

It is again trivial that

$$\lim_{\varepsilon\to 0} \limsup_{n\to\infty} P\{\Delta_{6n} > \eta\} = 0, \quad \eta > 0,$$

either by applying Lemma 2.4, or directly by Lemma 2.5 and (2.18). Hence Theorem 10.2 is proved.

THEOREM 10.3. Suppose that the density function $f = F'$ is positive on the open support of F. If

$$J(\tfrac{1}{\alpha}, \tfrac{1}{\beta}) = \sup_{0<u<1} \frac{u^{\alpha}(1-u)^{\beta}}{f(Q(u))} < \infty \quad \text{with} \quad 0 \le \alpha < \tfrac{3}{2}, \; 0 \le \beta < \tfrac{3}{2}, \tag{10.9}$$

then

$$\Delta_n^{(8)} = \sup_{0\le u\le 1} |g_n(u) - \Gamma_n(u)| \overset{a.s.}{=} O(n^{-\tau})$$

for any $\tau < \min(\tau(\alpha), \tau(\beta))$, where

$$\tau(\alpha) = \begin{cases} \frac{1}{4}, & \text{if } \alpha < 1/2, \\ \frac{1}{4\alpha+2}, & \text{if } 1/2 \le \alpha \le 1, \\ \frac{3-2\alpha}{10-4\alpha}, & \text{if } 1 \le \alpha < 3/2, \end{cases} \qquad \tau(\beta) = \begin{cases} \frac{1}{4}, & \text{if } \beta < 1/2, \\ \frac{3-2\beta}{10-4\beta}, & \text{if } 1/2 \le \beta \le 1, \\ \frac{3-2\beta}{6}, & \text{if } 1 \le \beta < 3/2. \end{cases}$$

Proof. Using (10.4)-(10.8) in the above proof of Theorem 10.2, we have

$$\Delta_n^{(8)} \le \Delta_n^{(2)} + \Delta_{1n} + \Delta_n^{(4)} + \Delta_{2n} + \tilde{\Delta}_{3n} + \ldots + \tilde{\Delta}_{6n}, \tag{10.10}$$

where $\tilde{\Delta}_{3n}, \ldots, \tilde{\Delta}_{6n}$ are defined as $\Delta_{3n}, \ldots, \Delta_{6n}$ in (10.8), respectively, but with n^{-1} replaced by $25n^{-1} \log\log n$. Here

$$\Delta_n^{(2)} \overset{a.s.}{=} O(n^{-\lambda}), \quad \lambda < \min(\tfrac{1}{2}, \tfrac{3}{2} - \beta), \tag{10.11}$$

by (5.3) and Lemma 3.3, and

$$\Delta_n^{(4)} \overset{a.s.}{=} O(n^{-\rho}), \quad \rho < \min(\tfrac{3-2\alpha}{10-4\alpha}, \tfrac{3-2\beta}{10-4\beta}), \tag{10.12}$$

by Lemma 5.3.

When estimating Δ_{1n} we have to distinguish three cases according to (5.6). If $\beta < 1$, then by (2.1)

$$\Delta_{1n} \stackrel{a.s.}{=} O(n^{-\frac{1}{2}}(\log n)^2).$$

If $\beta \geq 1$, then introducing $\varepsilon_{1n} = n^{-\tau_1}$, $0 < \tau_1 < 1$, we have

$$\begin{aligned}\Delta_{1n} &\leq \sup_{0\leq u\leq 1-\varepsilon_{1n}} |\alpha_n(u)-B_n(u)|Q(u)\\ &+ \sup_{1-\varepsilon_{1n}\leq u<1} |\alpha_n(u)|Q(u) + \sup_{1-\varepsilon_{1n}\leq u<1} |B_n(u)|Q(u) \qquad (10.13)\\ &= A_{1n}(\varepsilon_{1n}) + A_{2n}(\varepsilon_{1n}) + A_{3n}(\varepsilon_{1n}).\end{aligned}$$

By (2.1) and (5.6) we get

$$A_{1n}(\varepsilon_{1n}) \stackrel{a.s.}{=} \begin{cases} O(n^{-\frac{1}{2}}(\log n)^2 \log \varepsilon_{1n}^{-1}), & \text{if } \beta = 1,\\ O(n^{-\frac{1}{2}}(\log n)^2 \varepsilon_{1n}^{1-\beta}), & \text{if } \beta > 1,\end{cases}$$

and by (2.9) and with an arbitrarily small $\delta > 0$,

$$\begin{aligned}A_{3n}(\varepsilon_{1n}) &\stackrel{a.s.}{=} \begin{cases} (\log\log n)^{\frac{1}{2}} O(\sup_{1-\varepsilon_{1n}\leq y<1} (1-y)^{\frac{1}{2}-\delta}\log(1-y)), & \text{if } \beta = 1,\\ (\log\log n)^{\frac{1}{2}} O(\sup_{1-\varepsilon_{1n}\leq y<1} (1-y)^{\frac{1}{2}-\delta}(1-y)^{1-\beta}), & \text{if } \beta > 1,\end{cases} \qquad (10.14)\\ &= \begin{cases} O(\varepsilon_{1n}^{\frac{1}{2}-\delta}(\log \varepsilon_{1n})(\log\log n)^{\frac{1}{2}}), & \text{if } \beta = 1,\\ O(\varepsilon_{1n}^{\frac{3}{2}-\beta-\delta}(\log\log n)^{\frac{1}{2}}), & \text{if } \beta > 1.\end{cases}\end{aligned}$$

Using (2.8) instead of (2.9), we see that $A_{2n}(\varepsilon_{1n})$ is of exactly the same order as $A_{3n}(\varepsilon_{1n})$. Summing up, we obtain

$$\Delta_{1n} \stackrel{a.s.}{=} O(n^{-\lambda}), \quad \lambda < \min\left(\tfrac{1}{2}, \tfrac{3}{2} - \beta\right), \qquad (10.15)$$

since τ_1 may be taken to be arbitrarily close to 1.

Define $\varepsilon_{2n} = n^{-\tau_2}$ with

$$\tau_2 < \begin{cases} \frac{1}{2}, & \text{if } \alpha \leq \frac{1}{2},\\ \frac{1}{2\alpha+1}, & \text{if } \alpha > \frac{1}{2}\end{cases}$$

and

$$\varepsilon_{3n} = \begin{cases} (n^{-1}\log\log n)^{\frac{1}{2}}, & \text{if } \beta \le 1, \\ n^{-\tau_3}, \ \tau_3 < 1/3, & \text{if } \beta > 1. \end{cases}$$

Then for Δ_{2n} in (10.10) and (10.7),

$$\begin{aligned} \Delta_{2n} \le & \sup_{0 \le y < \varepsilon_{2n}} Q(y)|B_n(y)| + \sup_{0 \le y \le \varepsilon_{2n}} Q(U_n(y))|B_n(U_n(y))| \\ & + \sup_{\varepsilon_{2n} \le y \le 1/2} |Q(y)-Q(U_n(y))||B_n(y)| \\ & + \sup_{\varepsilon_{2n} \le y \le 1/2} |B_n(y)-B_n(U_n(y))|Q(U_n(y)) \\ & + \sup_{1/2 \le y \le 1-\varepsilon_{3n}} |Q(y)-Q(U_n(y))||B_n(y)| \\ & + \sup_{1/2 \le y \le 1-\varepsilon_{3n}} |B_n(y)-B_n(U_n(y))|Q(U_n(y)) \\ & + \sup_{1-\varepsilon_{3n} \le y \le 1} Q(y)|B_n(y)| + \sup_{1-\varepsilon_{3n} \le y \le 1} Q(U_n(y))|B_n(U_n(y))| \\ = \ & A_{4n} + \ldots + A_{11n} . \end{aligned} \tag{10.16}$$

So, let us start estimating them. By (2.11) and Borel-Cantelli,

$$A_{4n} \overset{a.s.}{=} O(n^{-\tau_2/2}(\log n)^{\frac{1}{2}}). \tag{10.17}$$

Now (2.14) implies that

$$\begin{aligned} A_{5n} & \le C_{18} \sup_{0 \le y \le U_n(\varepsilon_{2n})} |B_n(y)| \\ & \le C_{18} \sup_{0 < y \le \varepsilon_{2n} + (n^{-1}\log\log n)^{\frac{1}{2}}} |B_n(y)| \end{aligned}$$

almost surely for large enough (random) n , and hence by the argument leading to (10.17) we also have

$$A_{5n} \overset{a.s.}{=} O(n^{-\tau_2/2}(\log n)^{\frac{1}{2}}). \tag{10.18}$$

Further, by a one-term Taylor again, and using (2.14) and its counterpart (2.15) for B_n after,

$$A_{6n} = \sup_{\varepsilon_{2n} \le y \le 1/2} n^{-\frac{1}{2}} |u_n(y)| |B_n(y)| (f(Q(\tau_n(y))))^{-1}$$

$$(10.19) \quad = \sup_{\varepsilon_{2n} \le y \le 1/2} n^{-\frac{1}{2}} |u_n(y)| |B_n(y)| \frac{(\tau_n(y))^{\alpha}}{f(Q(\tau_n(y)))} \left(\frac{y}{\tau_n(y)}\right)^{\alpha} y^{-\alpha}$$

$$\stackrel{a.s.}{=} O(n^{-\frac{1}{2}} (\log\log n) \varepsilon_{2n}^{-\alpha})$$

by making use of condition (10.9) and the inequality in (6.24). The latter is possible since $\tau_n(y)$ satisfies (6.11) and $\varepsilon_{1n} \ge 25n^{-1} \log\log n$.

The next term is easy. Since $Q(U_n(y))$ is almost surely bounded on $[0,1/2]$, we obtain

$$(10.20) \quad A_{7n} \stackrel{a.s.}{=} O(n^{-1/4} (\log\log n)^{1/4} (\log n)^{1/2})$$

by applying first (2.14), and then (2.11) and Borel-Cantelli for the increments of order $(n^{-1} \log\log n)^{\frac{1}{2}}$ of the Brownian bridge. Proceeding now to the next term, we obtain

$$(10.21) \quad A_{8n} \stackrel{a.s.}{=} O(n^{-1/2} (\log\log n) \varepsilon_{3n}^{-\beta})$$

by a completely analogous procedure that led to the order of A_{6n} in (10.19).

By the argument yielding (10.20), we have also

$$A_{9n} \stackrel{a.s.}{=} O(n^{-1/4} (\log\log n)^{1/4} (\log n)^{1/2}) \sup_{1/2 \le y \le 1-\varepsilon_{3n}} Q(U_n(y)),$$

where, by (2.14), almost surely for large enough (random) n,

$$\sup_{1/2 \le y \le 1-\varepsilon_{3n}} Q(U_n(y)) \le Q(U_n(1-\varepsilon_{3n}))$$

$$\le Q(1-\varepsilon_{3n} + (n^{-1} \log\log n)^{1/2})$$

$$\stackrel{a.s.}{=} \begin{cases} O(1), & \text{if } \beta < 1, \\ O(\log n), & \text{if } \beta = 1, \\ O(\varepsilon_{3n}^{1-\beta}), & \text{if } \beta > 1, \end{cases}$$

on applying condition (10.9) via (5.6) and the fact that $\varepsilon_{3n} \ge (n^{-1} \log\log n)^{\frac{1}{2}}$. Altogether we have

$$(10.22) \quad A_{9n} \stackrel{a.s.}{=} O(n^{-\kappa_2}), \ \kappa_2 < \begin{cases} 1/4, & \text{if } \beta \le 1, \\ (3-2\beta)/6, & \text{if } \beta > 1. \end{cases}$$

Next we notice that $A_{10n} = A_{3n}(\varepsilon_{3n})$ in the notation of (10.13), and hence by (10.14),

$$(10.23)\qquad A_{10n} \overset{a.s.}{=} \begin{cases} O(\varepsilon_{3n}^{1/2}(\log n)^{1/2}), & \text{if } \beta < 1, \\ O(\varepsilon_{3n}^{\frac{1}{2}-\delta}(\log \varepsilon_{3n})(\log\log n)^{\frac{1}{2}}), & \text{if } \beta = 1, \\ O(\varepsilon^{\frac{3}{2}-\beta-\delta}(\log\log n)^{\frac{1}{2}}), & \text{if } \beta > 1, \end{cases}$$

since Q is bounded in the case $\beta < 1$, and the corresponding rate obtains from (2.11) directly. For A_{11n}, again by (2.14),

$$A_{11n} \le \sup_{U_n(1-\varepsilon_{3n}) \le y < 1} Q(y)\,|B_n(y)|$$

$$\le \sup_{1-\varepsilon_{3n}-(n^{-1}\log\log n)^{\frac{1}{2}} \le y \le 1} Q(y)\,|B_n(y)|$$

almost surely for large enough (random) n, and hence for A_{11n} we get the same formulae as in (10.23) with ε_{3n} replaced by $\varepsilon_{3n} + (n^{-1}\log\log n)^{\frac{1}{2}}$. Using this rate, the definitions of ε_{2n} and ε_{3n}, and collecting all the rates of convergence in (10.17)-(10.23), a simple computation gives

$$(10.24)\qquad \Delta_{2n} \overset{a.s.}{=} O(n^{-\kappa}), \quad \kappa = \min(\kappa_1, \kappa_2),$$

where κ_2 is as in (10.22), and

$$(10.25)\qquad \kappa_1 < \begin{cases} 1/4, & \text{if } \alpha \le 1/2, \\ \dfrac{1}{4\alpha+2}, & \text{if } \alpha > 1/2. \end{cases}$$

We proceed now to $\tilde{\Delta}_{3n}$ in (10.10), defined in (10.8), with the modification noted below (10.10), i.e., with n^{-1} replaced by

$$(10.26)\qquad \delta_n = 25n^{-1}\log\log n.$$

If $\beta \le 1$ then directly by the first two lines of (5.6) and (2.3),

$$\tilde{\Delta}_{3n} \overset{a.s.}{=} \begin{cases} O(n^{-1/4}(\log\log n)^{1/4}(\log n)^{1/2}), & \text{if } \beta < 1, \\ O(n^{-1/4}(\log\log n)^{1/4}(\log n)^{1/2}\log n), & \text{if } \beta = 1. \end{cases}$$

If $\beta > 1$, then set $\varepsilon_{4n} = n^{-\tau_4}$ with $\tau_4 < 1/2$. Again by (2.3) and (5.6) and further by (6.25) and (2.9),

$$\tilde{\Delta}_{3n} \le \sup_{0\le y\le 1-\varepsilon_{4n}} Q(y)\,|u_n(y)-B_n(y)|$$

$$+ \sup_{1-\varepsilon_{4n}\le y\le 1-\delta_n} Q(y)(1-y)^{\frac{1}{2}} \frac{|u_n(y)|}{(1-y)^{\frac{1}{2}}}$$

$$+ \sup_{1-\varepsilon_{4n}\le y\le 1-\delta_n} Q(y)(1-y)^{\frac{1}{2}-\delta} \frac{|B_n(y)|}{(1-y)^{\frac{1}{2}-\delta}}$$

$$\stackrel{a.s.}{=} O(n^{-1/4}(\log\log n)^{1/4}(\log n)^{1/2}\,\varepsilon_{4n}^{1-\beta})$$

$$+ O(\varepsilon_{4n}^{\frac{3}{2}-\beta}(\log\log n)^{\frac{1}{2}})$$

$$+ O(\varepsilon_{4n}^{\frac{3}{2}-\beta-\delta})$$

$$= O(\varepsilon_{4n}^{\frac{3}{2}-\beta-\delta})$$

with any small $\delta > 0$. Hence

(10.27) $$\tilde{\Delta}_{3n} \stackrel{a.s.}{=} O(n^{-\eta}), \quad \eta < \begin{cases} \frac{1}{4}, & \text{if } \beta \le 1, \\ \frac{3-2\beta}{4}, & \text{if } \beta > 1. \end{cases}$$

When estimating $\tilde{\Delta}_{4n}$ of (10.8) as modified in (10.10), we have to cut the interval $[0,1-\delta_n]$ into four pieces. Letting

$$\varepsilon_{5n} = \begin{cases} \delta_n, & \text{if } \alpha < 1, \\ n^{-\tau_5}, \ \tau_5 < \frac{1}{2\alpha-1}, & \text{if } \alpha \ge 1, \end{cases}$$

with δ_n as in (10.26), and

$$\varepsilon_{6n} = \begin{cases} \delta_n, & \text{if } \beta < 1, \\ n^{-\tau_6}, \ \tau_6 < 1, & \text{if } \beta \ge 1, \end{cases}$$

we have

$$\tilde{\Delta}_{4n} \le \sup_{0\le y\le \varepsilon_{5n}} Q(y)\,|u_n(y)|$$

$$+ \sup_{0\le y\le \varepsilon_{5n}} Q(\tau_n(y))\,|u_n(y)|$$

$$+ \sup_{\varepsilon_{5n} \le y \le 1/2} |Q(\tau_n(y)) - Q(y)|\,|u_n(y)|$$

(10.28) $$+ \sup_{1/2 \le y \le 1-\varepsilon_{6n}} |Q(\tau_n(y)) - Q(y)|\,|u_n(y)|$$

$$+ \sup_{1-\varepsilon_{6n} \le y \le 1-\delta_n} Q(y)\,|u_n(y)|$$

$$+ \sup_{1-\varepsilon_{6n} \le y \le 1-\delta_n} Q(\tau_n(y))\,|u_n(y)|$$

$$= A_{12n} + \ldots + A_{17n} .$$

Since $Q(\tau_n(y))$ is eventually bounded on $[0,\varepsilon_{5n}]$, (2.12) gives

(10.29) $$A_{12n} + A_{13n} \overset{a.s.}{=} O(\max(n^{-\frac{1}{2}}\log n,\ (\varepsilon_{5n} \log n)^{\frac{1}{2}})).$$

By a one-term Taylor expansion

$$A_{14n} = \sup_{\varepsilon_{5n} \le y \le 1/2} \frac{|u_n(y)|\,|\tau_n(y)-y|}{f(Q)(\tau_n^*(y)))}$$

(10.30) $$\le n^{-\frac{1}{2}} \sup_{\varepsilon_{5n} \le y \le 1/2} \frac{u_n^2(y)}{f(Q(\tau_n^*(y)))}$$

$$\le n^{-\frac{1}{2}} \sup_{\varepsilon_{5n} \le y \le 1/2} \frac{(\tau_n^*(y))^{\alpha}}{f(Q(\tau_n^*(y)))} \frac{u_n^2(y)}{y} \left(\frac{y}{\tau_n^*(y)}\right)^{\alpha} y^{1-\alpha},$$

by (6.11), and by (6.11) $\tau_n^*(y)$ also satisfies the inequalities

$$y \wedge U_n(y) \le \tau_n^*(y) \le y \vee U_n(y).$$

The latter, via Lemma 2.9, imply that

(10.31) $$\limsup_{n\to\infty} \sup_{\delta_n \le y \le 1-\delta_n} \frac{y(1-y)}{\tau_n^*(y)(1-\tau_n^*(y))} \le 36 \quad \text{a.s.}$$

Hence by condition (10.9) and (6.25),

(10.32) $$A_{14n} \overset{a.s.}{=} \begin{cases} O(n^{-\frac{1}{2}}\log\log n) & , \text{ if } \alpha < 1, \\ O(n^{-\frac{1}{2}}(\log\log n)\varepsilon_{5n}^{1-\alpha}), & \text{if } \alpha \ge 1. \end{cases}$$

Upon replacing y by $1-y$, $\tau_n^*(y)$ by $1-\tau_n^*(y)$ and α by β in the third line of (10.30), we obtain

$$(10.33) \qquad A_{15n} \overset{a.s.}{=} \begin{cases} O(n^{-\frac{1}{2}}\log\log n), & \text{if } \beta < 1, \\ O(n^{-\frac{1}{2}}(\log\log n)\varepsilon_{6n}^{1-\beta}), & \text{if } \beta \geq 1. \end{cases}$$

Condition (10.9) via (5.6) and (2.13) gives

$$(10.34) \qquad A_{16n} \overset{a.s.}{=} \begin{cases} O(n^{-\frac{1}{2}}\log n), & \text{if } \beta < 1, \\ O(n^{-\frac{1}{2}}(\log n)^2), & \text{if } \beta = 1, \\ O(\varepsilon_{6n}^{\frac{1}{2}}(\log n)^{\frac{1}{2}}\delta_n^{1-\beta}), & \text{if } \beta > 1. \end{cases}$$

For the last term in (10.28) we clearly have

$$A_{17n} \leq Q(\tau_n(1-\delta_n)) \sup_{1-\varepsilon_{6n} \leq y \leq 1} |u_n(y)|,$$

where Lemma 2.9 again gives

$$\limsup_{n\to\infty} \frac{1-\tau_n(1-\delta_n)}{\delta_n} \geq \frac{1}{6} \quad \text{a.s.}$$

Hence (5.6) and (2.13) together give that $A_{17n} \overset{a.s.}{=} O(A_{16n})$, and (10.28), (10.29), (10.32)-(10.34) imply through some computations that

$$(10.35) \quad \tilde{\Delta}_{4n} \overset{a.s.}{=} O(n^{-\nu}), \quad \nu < \begin{cases} \min(\frac{1}{2}, \frac{3}{2} - \beta), & \text{if } \alpha < 1, \\ \min(\frac{1}{4\alpha-2}, \frac{3}{2} - \beta), & \text{if } \alpha \geq 1. \end{cases}$$

Again (5.6) and (2.9) imply that

$$(10.36) \qquad \tilde{\Delta}_{5n} \overset{a.s.}{=} \begin{cases} O(n^{-\frac{1}{2}+\delta}), & \text{if } \beta < 1, \\ O(n^{-(\frac{3}{2}-\beta)+\delta}), & \text{if } \beta \geq 1, \end{cases}$$

with $\delta > 0$ arbitrarily small.

Finally,

$$\begin{aligned} \tilde{\Delta}_{6n} &\leq \max\{Q(1-n^{-1}), Q(1-U_{n:n})\} \sup_{1-\delta_n \leq y \leq 1} |u_n(y)| \\ &\quad + n^{\frac{1}{2}} \int_{1-n^{-1}}^{1} Q(x)\,dx \\ &\quad + n^{\frac{1}{2}} \int_{U_{n:n}}^{1} Q(x)\,dx \\ &= A_{18n} + A_{19n} + A_{20n}. \end{aligned}$$

Since

$$\liminf_{n\to\infty} n(\log n)^2(1-U_{n:n}) \geq C_{19} > 0$$

and

$$\liminf_{n\to\infty} n(\log n)^{-1}(1-U_{n:n}) \leq C_{20} < \infty ,$$

as stated in Lemma 2.10, and

$$\sup_{1-\delta_n\leq y\leq 1} |u_n(y)| \overset{a.s.}{=} O(n^{-\frac{1}{2}} \log n),$$

(5.6) and a simple computation yields

$$A_{18n} \overset{a.s.}{=} \begin{cases} O(n^{-\frac{1}{2}} \log n) & , \text{ if } \beta < 1, \\ O(n^{-\frac{1}{2}}(\log n)^2) & , \text{ if } \beta = 1, \\ O(n^{-\frac{1}{2}}(\log n)\{n(\log n)^2\}^{\beta-1}), & \text{ if } \beta > 1, \end{cases}$$

$$A_{19n}+A_{20n} \overset{a.s.}{=} \begin{cases} O(n^{-\frac{1}{2}} \log n) & , \text{ if } \beta < 1, \\ O(n^{-\frac{1}{2}+\delta}(\log n)^{1-\delta}), & \text{ if } \beta = 1, \\ O((n^{-1}\log n)^{\frac{3}{2}-\beta}) & , \text{ if } \beta > 1. \end{cases}$$

Thus

$$(10.37) \quad \tilde{\Delta}_{6n} \overset{a.s.}{=} \begin{cases} O(n^{-\frac{1}{2}} \log n) & , \text{ if } \beta < 1 \\ O(n^{-\frac{1}{2}+\delta}) & , \text{ if } \beta = 1 \\ O(n^{-(\frac{3}{2}-\beta)+\delta}) & , \text{ if } \beta > 1. \end{cases}$$

with an arbitrarily small $\delta > 0$.

Collecting now the respective rates of convergence in (10.11), (10.12), (10.15), (10.24), (10.25), (10.27), (10.35)-(10.37) and comparing them, Theorem 10.3 follows.

11. EMPIRICAL LORENZ PROCESSES.

We recall the definitions of the theoretical and empirical Lorenz curves L_F and L_n in (1.8) and (1.10) respectively, together with that of the empirical Lorenz process ℓ_n in (1.13). Theorem 10.1 and the strong law of large numbers readily imply the strong uniform consistency:

THEOREM 11.1. *If* $\mu < \infty$ *then*

$$\sup_{0\leq y\leq 1} |L_n(y) - L_F(y)| \xrightarrow{a.s.} 0.$$

Introduce now the following sequence of mean-zero Gaussian processes, the members of which are the same in distribution for each n:

$$\begin{aligned}\Lambda_n(y) &= \mu^{-1}\Gamma_n(y) - \mu^{-2}G_F(y)\Gamma_n(1)\\ &= \mu^{-1}\{\Gamma_n(y)-L_F(y)\Gamma_n(1)\}, \quad 0 \leq y \leq 1,\end{aligned}$$

where $\Gamma_n(\cdot)$ is that of Theorem 10.2. Note that $\Lambda_n(\cdot)$ results from $\Gamma_n(\cdot)$, the approximating Gaussian process of the unscaled empirical Lorenz process $g_n(\cdot)$, by a transformation of the same structure that gave the approximating Gaussian process $S_n(\cdot)$ of the scaled total time on test process from $T_n(\cdot)$ of the unscaled total time on test process (cf. Theorem 7.2). Since

$$\begin{aligned}\ell_n(y) &= \mu^{-1}g_n(y) - \mu^{-2}G_F(y)g_n(1)\\ &\quad + g_n(y)\left\{\frac{1}{G_n(1)} - \frac{1}{G_F(1)}\right\}\\ &\quad + \frac{G_F(y)}{\mu} g_n(1)\left\{\frac{1}{G_F(1)} - \frac{1}{G_n(1)}\right\},\end{aligned}$$

analogously to the just mentioned case (cf. the proof of Theorem 7.2), Theorems 10.1 and 10.2 imply the following result.

THEOREM 11.2. *If the conditions of Theorem 10.2 are satisfied, then*

$$\sup_{0\leq y\leq 1} |\ell_n(y) - \Lambda_n(y)| \xrightarrow{P} 0. \tag{11.1}$$

Similarly, Theorem 10.3 and its loglog law consequence imply the next theorem just as Theorem 6.3 implied Theorem 7.3.

THEOREM 11.3. *If the conditions of Theorem 10.3 are satisfied, then*

$$\sup_{0\le y\le 1} |\ell_n(y) - \Lambda_n(y)| \overset{a.s.}{=} O(n^{-\tau})$$

<u>with the same</u> τ <u>as in Theorem 10.3.</u>

Before discussing these results, we note that Theorem 11.1 was first proved by Goldie (1977). In fact he proved uniform consistency results without postulating the continuity of F (which is assumed in the present work throughout). As he noted, in this case the different definitions of the theoretical and empirical Lorenz curves do come into play and the limits in his two results are different.

It was again Goldie (1977) who first proved the statement of Theorem 11.2 in his Theorem 6 by different methods, assuming the continuity of Q, $EX^2 < \infty$ and an extra variation condition. He gives five equivalent forms of his variation condition in his Proposition 7, one of which is the following: There exist positive constants C, $a < 1$ and $u_o < 1$ such that

$$\frac{Q(s)}{Q(t)} \le C\left(\frac{1-t}{1-s}\right)^a \quad \text{whenever} \quad u_o < t \le s < 1. \tag{11.2}$$

Goldie (1977) also proved the weak convergence of $\ell_n(\cdot)$ in Skorohod's M_1 topology to the same limit process $\Lambda(\cdot)$ as above in the case when Q is possibly discontinuous and under the just mentioned conditions, i.e., under the finiteness of the second moment, the continuity of F and the variation condition of (11.2). Subsequently, Sendler (1982) also proved the weak convergence of $\ell_n(\cdot)$ to $\Lambda(\cdot)$ relative to the supremum and M_1 topologies, according to whether Q is continuous or not, under the condition that $EX^{2+\delta} < \infty$. He obtains his result for $\ell_n(\cdot)$ as a special case of a more general result on "functionals of order statistics".

For the approximating processes Λ_n one can easily prove a log log law and this is then inherited by ℓ_n.

COROLLARY 11.4. <u>Under the conditions of Theorem 11.3</u>

$$\limsup_{n\to\infty} \left(\frac{n}{\log\log n}\right)^{1/2} \sup_{0\le u\le 1} |L_n(u)-L_F(u)| \le \frac{4}{\mu}\int_0^1 h(y)\,dQ(y)$$

almost surely, where $h(y) = (y(1-y)\log\log \frac{1}{y(1-y)})^{1/2}$, $y \in (0,1)$.

12. DISCUSSION OF RESULTS ON EMPIRICAL LORENZ PROCESSES

1) The limit process of the unscaled Lorenz process $g_n(u) = n^{\frac{1}{2}}(G_n(u)-G_F(u))$, $0 \leq u \leq 1$, is the mean-zero Gaussian process

$$\Gamma(u) = \Gamma_F(u) = \int_0^u B(y)\,dQ(y), \quad 0 \leq u \leq 1.$$

If $F_{\theta,\sigma}(x) = F((x-\theta)/\sigma)$, $-\infty < \theta < \infty$, $\sigma > 0$, then

$$\Gamma_{F_{\theta,\sigma}}(u) \equiv \sigma\Gamma_F(u),$$

that is, Γ_F is shift-free. The covariance function of Γ_F is

$$\sigma_3(s,t) = E\Gamma_F(s)\Gamma_F(t) = \int_0^s \int_0^t (\min(u,v)-uv)\,dQ(u)\,dQ(v)$$

for $0 \leq s, t < 1$. If $0 \leq s \leq t \leq 1$ we obtain

$$\sigma_3(s,t) = 2\int_0^s (1-u)\{Q(u)-N_F^1(u)\}dQ(u)$$

$$+ \{N_F^1(t)-N_F^1(s)\}\{Q(s)-N_F^1(s)\},$$

where

$$N_F^1(u) = H_F^{-1}(u)-t_F = \int_0^u (1-y)\,dQ(y),$$

and hence the variance function is

$$\sigma_3^2(t) = 2\int_0^t (1-u)\{Q(u)-N_F^1(u)\}dQ(u).$$

2) The limit process of the empirical Lorenz process $\ell_n(u) = n^{\frac{1}{2}}(L_n(u)-L_F(u))$, $0 \leq u \leq 1$, is the mean-zero Gaussian process

$$\Lambda(u) = \Lambda_F(u) = \frac{1}{\mu}\{\int_0^u B(y)\,dQ(y)-L_F(u)\int_0^1 B(y)\,dQ(y)\},$$

$0 \leq u \leq 1$. If $F_\sigma(x) = F_{0,\sigma}(x) = F(x/\sigma)$, $\sigma > 0$, then

$$\Lambda_{F_\sigma}(u) = \Lambda_F(u), \quad 0 \leq u \leq 1,$$

i.e., Λ_F is scale-free. The covariance function of Λ_F is

$$\sigma_4(s,t) = E\Lambda_F(s)\Lambda_F(t)$$

$$= \mu^{-2}\{\sigma_3(s,t) + L_F(s)L_F(t)\sigma_3(1,1)$$

$$- L_F(t)\sigma_3(s,1) - L_F(s)\sigma_3(t,1)\}.$$

Although Goldie's (1977) representation of the limit process Λ_F

is the same as ours, he gives the covariance function in a more complicated form in terms of the truncated second-moment function

$$\int_0^{Q(t)} x^2 dF(x)$$

arising in a natural way in his proof of the tightness of the sequence $\{\ell_n(\cdot)\}$.

The variance function is

$$\begin{aligned}\sigma_4^2(t) &= E\Lambda_F^2(t)\\ &= \frac{1}{\mu^2}\{2\int_0^t (1-u)\{Q(u)-N_F^1(u)\}dQ(u)[1-2L_F(t)]\\ &+ 2L_F^2(t)\int_0^1 (1-u)\{Q(u)-N_F^1(u)\}dQ(u)\\ &-2L_F(t)[N_F^1(1)-N_F^1(t)][Q(t)-N_F^1(t)]\}.\end{aligned}$$

3) Since we could not identify Γ_F or Λ_F as a known process for any specified F and could not compute the distribution of any of their functionals, we must use the bootstrap method of Section 17. We can, of course, again draw consequences of the convergence theory of Lorenze processes pointwise. Let us choose a fixed point $u \in (0,1)$ and consider the estimator $\sigma_{4n}^2(u)$ of the limiting variance $\sigma_4^2(u)$ obtained by replacing μ by the sample mean $\bar{X}_n$, L_F by the empirical Lorenz curve L_n, Q by the sample quantile function Q_n in (8.5) and N_F^1 by its empirical counterpart

$$N_n^1(u) = \int_0^u (1-y)\,dQ_n(y)$$

in the definition of $\sigma_4^2(u)$ above. Note that $N_n^1(1) = \bar{X}_n - X_{1:n}$

THEOREM 12.1. (i) <u>If</u> $\mu < \infty$ <u>then</u>

$$L_n(u) \xrightarrow{a.s.} L_F(u).$$

(ii) <u>If</u> Q <u>is continuous at</u> u <u>and</u> $EX^2 < \infty$, <u>then</u>

$$\lim_{n\to\infty} pr\{\frac{\ell_n(u)}{\sigma_4(u)} \le x\} = \Phi(x), \quad -\infty < x < \infty.$$

(iii) <u>If</u> $EX^2 < \infty$ <u>and</u> Q <u>is continuous on</u> $[0,u+\varepsilon)$ <u>with any small</u> $\varepsilon > 0$, <u>then</u>

$$\lim_{n\to\infty} pr\{L_n(u) - x\frac{\sigma_{4n}(u)}{\sqrt{n}} < L_F(u) < L_n(u) + x\frac{\sigma_{4n}(u)}{\sqrt{n}}\} = 2\Phi(x)-1$$

for any x *on the line*.

Proof. The first two parts follow from simple inspections of the proofs of Theorems 10.1 and 10.2. Part (iii) follows from (ii) upon noting that $\sigma_{4n}(u) \to \sigma_4(u)$ almost surely under the stated condition. The latter claim follows from the corresponding version of Theorem 10.1.

Chandra and Singpurvalla (1978) proved Part (ii) of Theorem 12.1 under the stronger assumption that Q has a nonzero continuous derivative at u and $\mu < \infty$. They claim the weak convergence of $\ell_n(\cdot)$ to $\Lambda_F(\cdot)$ assuming the just mentioned condition on Q at each $u \in (0,1)$ but they have a "pointwise proof" only. Indeed, their claim is clearly not true since $\Lambda_F(\cdot)$ blows up if $EX^2 = \infty$, i.e., $\sup\{|\Lambda_F(u)| : 0 \leq u \leq 1\} = \infty$ almost surely. This follows immediately from the representation in (10.3), for example. Gail and Gastwirth (1978a,b) applied $\ell_n(u)/\sigma_4(u)$, in particular with $u = 1/2$, as a test statistic for testing scale-free exponentiality. Sendler (1979) gives similar results to that in Part (iii) of Theorem 12.1. Besides the exponential distribution he computes $\sigma_4^2(u)$ for the rectangular distribution on $[0,1]$ and for the Pareto family of distribution functions $F(x) = 1-x^{-\beta}$, $x \geq 1$.

13. THE EMPIRICAL CONCENTRATION PROCESS OF GOLDIE

Since the theoretical Lorenz curve

$$L_F(u) = \mu^{-1} \int_0^u Q(y)\,dy$$

is continuous and strictly increasing on [0,1], it has a well-defined continuous and strictly increasing inverse L_F^{-1} on [0,1] which we call the concentration curve pertaining to F. Since our F is always assumed to be continuous, Goldie's (1977) formula for L_F^{-1},

$$L_F^{-1}(u) = u \int_0^u \{1/\tilde{G}^{-1}(y)\}dy, \quad 0 \leq u \leq 1$$

is valid in our case, i.e., there is no problem with our convention of defining inverse functions to be also right-continuous, where $\tilde{G}^{-1}$ is the inverse function to the first moment distribution function of F:

$$\tilde{G}(t) = \mu^{-1} \int_0^t x\,dF(x).$$

The inverse empirical Lorenz curve

$$L_n^{-1}(u) = \inf\{y : L_n(y) > u\},$$

or what we call the empirical concentration curve may be described in more detail as

$$L_n^{-1}(u) = \begin{cases} 0, & \text{if } 0 \leq u \leq (n\bar{X}_n)^{-1} X_{1:n} \\ \dfrac{k-1}{n}, & \text{if } (n\bar{X}_n)^{-1} \sum_{i=1}^{k-1} X_{i:n} \leq u < (n\bar{X}_n)^{-1} \sum_{i=1}^{k} X_{i:n}, \quad 2 \leq k \leq n, \\ 1, & \text{if } u = 1. \end{cases}$$

Introducing

$$C_n(u) = (\bar{X}_n)^{-1} \int_0^u Q(x-)\,dE_n(x), \tag{13.1}$$

our observation in (10.1) gives that

$$P\{\sup_{0 \leq y \leq 1} |L_n(y) - C_n(U_n(y))| = 0\} = 1.$$

Since the inverse function to the compound function $C_n(U_n(\cdot))$ is $E_n(C_n^{-1}(\cdot))$, this implies that

$$P\{\sup_{0 \leq u \leq 1} |L_n^{-1}(u) - E_n(C_n^{-1}(u))| = 0\} = 1. \tag{13.2}$$

The following lemma will be basic for all the considerations in this section. It is in a sense a special case of a more general Lemma of Horváth (1984a,b).

LEMMA 13.1. On each $\omega \in \Omega$,

$$\sup_{0\le u\le 1} |L_n^{-1}(L_F(u))-u| = \sup_{0\le u\le 1} |L_F^{-1}(L_n(u))-u|.$$

Proof. The compound functions on both sides are step functions on $[0,1]$. The points where $L_n^{-1}(L_F(\cdot))$ jumps are

$$u_k = L_F^{-1}((n\bar{X}_n)^{-1}\sum_{i=1}^{k} X_{i:n}),\ 1=1,\dots,n,$$

where, of course, $u_n = 1$. The supremum on the left hand side is

$$\max_{1\le k\le n} \max(|L_n^{-1}(L_F(u_k))-u_k|, |L_n^{-1}(L_F(u_k-))-u_k|).$$

Clearly

$$|L_n^{-1}(L_F(u_k))-u_k| = |L_F^{-1}(L_n(\tfrac{k}{n}-)) - \tfrac{k}{n}|$$

and

$$|L_n^{-1}(L_F(u_k-))-u_k| = |L_F^{-1}(L_n(\tfrac{k-1}{n})) - \tfrac{k-1}{n}|,$$

since the arising rectangular triangles with hypotenuses on the line $y = x$ have equal perpendicular sides. But the right hand side supremum above is

$$\max_{1\le k\le n} \max(|L_F^{-1}(L_n(\tfrac{k}{n}))-\tfrac{k}{n}|, |L_F^{-1}(L_n(\tfrac{k}{n}-))-\tfrac{k}{n}|)$$

and hence the lemma.

Now the consistency result will follow easily. Our approach to proving the uniform convergence of inverse functions is perhaps the first alternative Goldie (1977) mentions before stating his corresponding Theorem 2.

THEOREM 13.2. If $\mu < \infty$ then

$$\Delta_n^{(9)} = \sup_{0\le u\le 1} |L_n^{-1}(u) - L_F^{-1}(u)| \xrightarrow{a.s.} 0.$$

Proof. By the fact that

(13.3) $\qquad L_F : [0,1] \to [0,1]$ is a homeomorphism

and by Lemma 13.1 we have

$$\Delta_n^{(9)} = \sup_{0\le u\le 1} |L_n^{-1}(L_F(u))-L_F^{-1}(L_F(u))|$$

$$= \sup_{0\le u\le 1} |L_F^{-1}(L_n(u))-L_F^{-1}(L_F(u))|,$$

and the latter goes to zero almost surely by Theorem 11.1 and the continuity of L_F^{-1}.

Since by (13.2)

$$\sup_{0\le u\le 1} |L_n^{-1}(u)-C_n^{-1}(u)| \le \sup_{0\le u\le 1} |E_n(u)-u|,$$

Theorem 13.2 of course implies that

$$\sup_{0\le u\le 1} |C_n^{-1}(u)-L_F^{-1}(u)| \xrightarrow{a.s.} 0 \tag{13.4}$$

if the mean is finite, which is naturally always assumed.

When approximating the Goldie concentration process $c_n(\cdot) = n^{\frac{1}{2}}(L_n^{-1}(\cdot)-L_F^{-1}(\cdot))$, the process

$$\ell_n^*(u) = n^{\frac{1}{2}}(C_n(u)-L_F(u)),\ 0 \le u \le 1 \tag{13.5}$$

will play a decisive role. We introduce the mean-zero Gaussian processes

$$\Gamma_n^*(u) = \frac{1}{\mu}\{L_F(u)\int_0^1 Q(y)dB_n(y) - \int_0^u Q(y)dB_n(y)\}. \tag{13.6}$$

LEMMA 13.3. *If* $EX^2 < \infty$ *and* Q *is continuous on* $[0,1]$, *then*

$$\Delta_n^{(10)} = \sup_{0\le u\le 1} |\ell_n^*(u)-\Gamma_n^*(u)| \xrightarrow{P} 0.$$

Proof. Since

$$\ell_n^*(u) = \frac{1}{\bar{X}_n}\{L_F(u)\int_0^1 Q(y)d\alpha_n(y) - \int_0^u Q(y)d\alpha_n(y)\}, \tag{13.7}$$

$$\Delta_n^{(10)} \le \frac{2}{\bar{X}_n} \sup_{0\le u\le 1} |\int_0^u Q(y)d(\alpha_n(y)-B_n(y))|$$

$$+ |\bar{X}_n^{-1}-\mu^{-1}| \sup_{0\le u\le 1} |\Gamma_n^*(u)|,$$

where the second term goes to zero by the law of large numbers and by (3.4) or Lemma 5.1. The supremum in the first term is not greater than

$$\sup_{0\le u\le 1} |\int_0^u (\alpha_n(y)-B_n(y))dQ(y)|$$

$$+ \sup_{0\le u\le 1} |Q(u)(\alpha_n(u)-B_n(u))|,$$

and the latter two random variables go to zero in probability by Lemmas 3.2, 2.4 and 2.5.

LEMMA 13.4. *If* $EX^2 < \infty$ *and* Q *is continuous on* $[0,1)$, *then*

$$\limsup_{n\to\infty} \sup_{0\le u\le\varepsilon} \frac{|\ell_n^*(u)|}{Q(u)q(u)} \xrightarrow{P} 0 \quad \text{as} \quad \varepsilon \downarrow 0$$

<u>for any O'Reilly weight function</u> q.

<u>Proof</u>. Since

(13.8) $$L_F(u) \le \frac{1}{\mu} uQ(u), \quad 0 \le u \le 1,$$

$$\sup_{0\le u\le\varepsilon} \left|\frac{\ell_n^*(u)}{Q(u)q(u)}\right| \le \frac{1}{\mu\bar{X}_n} \int_0^1 Q(y)\,d\alpha_n(y) \sup_{0\le u\le\varepsilon} \frac{u}{q(u)}$$

$$+ \frac{1}{\bar{X}_n} \sup_{0\le u\le\varepsilon} \left|\frac{\alpha_n(u)}{q(u)}\right| \sup_{0\le u\le\varepsilon} \left(\frac{Q(u)-Q(0)}{Q(u)} + 1\right)$$

by (13.7) and integration by parts. Hence the lemma follows by Lemma 3.2, Lemma 2.1, the law of large numbers and Lemma 2.6.

Let us consider now the Gaussian processes

$$\Psi_n(y) = -\frac{dL_F^{-1}(y)}{dy} \Lambda_n(L_F^{-1}(y))$$

(13.9) $$= -\frac{\mu}{Q(L_F^{-1}(y))} \Lambda_n(L_F^{-1}(y))$$

$$= \Phi_n(L_F^{-1}(y)), \quad 0 \le y \le 1,$$

where Λ_n is the approximating process of Section 11 to the empirical Lorenz process ℓ_n. Assuming the continuity of Q and the existence of the second moment (cf.(5.1) in the proof of Lemma 5.1) we may write, upon integrating by parts,

$$\Phi_n(u) = \frac{1}{Q(u)} \{L_F(u) \int_0^1 B_n(y)\,dQ(y) - \int_0^u B_n(y)\,dQ(y)\}$$

(13.10) $$= \frac{1}{Q(u)}\{-B_n(u)Q(u) + \int_0^u Q(y)\,dB_n(y) - L_F(u)\int_0^1 Q(y)\,dB_n(y)\}$$

$$= -\frac{\mu}{Q(u)} \Gamma_n^*(u) - B_n(u), \quad 0 \le u \le 1,$$

where $\Gamma_n^*(u)$ is of (13.6). We are now ready to state the weak approximation result for the Goldie concentration process $c_n(u) = n^{1/2}(L_n^{-1}(u) - L_F^{-1}(u))$.

THEOREM 13.5. <u>If</u> $EX^2 < \infty$, Q <u>is continuous on</u> $[0,1)$ <u>and for each</u> $\lambda > 1$

(13.11) $$\limsup_{u\to 0} \frac{Q(u)q(u)}{Q(u/\lambda)} < \infty$$

holds for some O'Reilly weight function q, then

$$\Delta_n^{(11)} = \sup_{0\le y\le 1} |c_n(y)-\Psi_n(y)| \xrightarrow{P} 0.$$

Proof. On the basis of the representation in (13.2)

$$c_n(u) = n^{\frac12}(E_n(C_n^{-1}(u))-L_F^{-1}(u))$$
$$= -\alpha_n(C_n^{-1}(u)) + n^{\frac12}(C_n^{-1}(u)-L_F^{-1}(u)),$$

and hence, by (13.10),

$$\Delta_n^{(11)} \le \sup_{0\le u\le 1} |B_n(C_n^{-1}(u))-\alpha_n(C_n^{-1}(u))|$$
$$+ \sup_{0\le u\le 1} |B_n(L_F^{-1}(u))-B_n(C_n^{-1}(u))|$$
$$+ \sup_{0\le u\le 1} |n^{\frac12}\{C_n^{-1}(u)-L_F^{-1}(u)\}+\frac{\mu}{Q(L_F^{-1}(u))}\Gamma_n^*(L_F^{-1}(u))|.$$

The first term is not greater than the left hand side of (2.1) and hence goes to zero almost surely. The second term goes to zero in probability by (13.4) and the continuity of the paths of a Brownian bridge. By (13.4) the third term is

$$A_{1n} = \sup_{0\le u\le 1} |n^{\frac12}\{C_n^{-1}(L_F(u))-u\} + \frac{\mu}{Q(u)}\Gamma_n^*(u)|.$$

We write the empirical process figuring here as

$$n^{\frac12}\{C_n^{-1}(L_F(u))-u\} = -n^{\frac12}\{L_F^{-1}(C_n(C_n^{-1}(L_F(u))))-C_n^{-1}(L_F(u))\}$$
$$+ n^{\frac12}\{L_F^{-1}(C_n(C_n^{-1}(L_F(u))))-u\},$$

and introduce

$$\gamma_n(u) = n^{\frac12}\{L_F^{-1}(C_n(u))-u\}$$
$$= n^{\frac12}\{L_F^{-1}(C_n(u))-L_F^{-1}(L_F(u))\},\quad 0\le u\le 1.$$

Thus

$$n^{\frac12}\{C_n^{-1}(L_F(u))-u\} = -\gamma_n(C_n^{-1}(L_F(u))) + n^{\frac12}\{L_F^{-1}(C_n(C_n^{-1}(L_F(u))))-u\}. \tag{13.12}$$

First we prove that

$$A_{2n} = \sup_{0\le u\le 1} |\frac{\mu}{Q(u)}\Gamma_n^*(u)-\gamma_n(u)| \xrightarrow{P} 0 \tag{13.13}$$

and then show that the time scale distortion $C_n^{-1}(L_F(\cdot))$ in the argument of γ_n does not change (13.13), and finally we shall show that the second factor in (13.12) is negligible.

To show (13.13) let $\varepsilon \in (0,1)$ be given and $n_o = n_o(\omega)$ be chosen so that $U_{1:n}(\omega) \leq \varepsilon$ if $n \geq n_o$. For such n's,

$$
\begin{aligned}
A_{2n} \leq & \sup_{0\leq u\leq U_{1:n}} |\gamma_n(u)| + \sup_{U_{1:n}\leq u\leq \varepsilon} |\gamma_n(u)| \\
& + \sup_{0\leq u\leq \varepsilon} \left|\frac{\mu}{Q(u)} \Gamma_n^*(u)\right| \\
& + \sup_{\varepsilon\leq u\leq 1} \left|\frac{\mu}{Q(u)} \Gamma_n^*(u) - \gamma_n(u)\right|
\end{aligned}
\tag{13.14}
$$

$$= A_{3n} + A_{4n}(\varepsilon) + A_{5n}(\varepsilon) + A_{6n}(\varepsilon).$$

Since $C_n(u) = 0$ on $[0,U_{1:n})$, $A_{3n} = n^{\frac{1}{2}}U_{1:n}$ converges to zero in probability. To handle A_{5n} we first note that by Lemma 5.1 the process

$$\frac{\mu}{Q(u)} \Gamma_n^*(u) = \frac{L_F(u)}{Q(u)} \int_0^1 Q(y)\,dB_n(y) - \frac{1}{Q(u)} \int_0^u Q(y)\,dB_n(y)$$

is almost surely continuous on $(0,1]$. Now we show that

$$\lim_{u\to 0} \frac{\mu}{Q(u)} \Gamma_n^*(t) = 0 \quad \text{a.s.}$$

This is true for the first term of the process by (13.8), while the modulus of the second term of it is not greater than

$$\left(\frac{Q(u)-Q(0)}{Q(u)} + 1\right) \sup_{0\leq t\leq u} |B_n(t)| \xrightarrow{\text{a.s.}} 0, \text{ as } u \downarrow 0,$$

obtained by partial integration. Hence

$$\limsup_{n\to\infty} A_{5n}(\varepsilon) \xrightarrow{P} 0 \quad \text{as} \quad \varepsilon \downarrow 0.$$

Next,

$$
\begin{aligned}
A_{6n}(\varepsilon) & \leq \sup_{\varepsilon\leq u\leq 1} \frac{\mu}{Q(L_F^{-1}(\tau_n(u))} \sup_{\varepsilon\leq u\leq 1} |\ell_n^*(u) - \Gamma_n^*(u)| \\
& \leq \sup_{\varepsilon\leq u\leq 1} \mu|\Gamma_n^*(u)| \left|\frac{1}{Q(u)} - \frac{1}{Q(L_F^{-1}(\tau_n(u)))}\right|
\end{aligned}
$$

by a one-term Taylor expansion, where ℓ_n^* is of (13.5) and

$$\min(C_n(u), L_F(u)) \leq \tau_n(u) \leq \max(C_n(u), L_F(u)). \tag{13.15}$$

Now Theorem 11.1 implies that

(13.16) $$\sup_{0\le u\le 1} |C_n(u)-L_F(u)| \xrightarrow{a.s.} 0.$$

Therefore, in view of the fact that $Q(\varepsilon) > 0$, Lemma 13.3 implies that

$$A_{6n}(\varepsilon) \xrightarrow{P} 0 \quad \text{for each fixed} \quad \varepsilon > 0.$$

So, in order to prove (13.13), there remains only the hardest term $A_{4n}(\varepsilon)$ to be estimated.

Again by Taylor,

(13.17) $$A_{4n}(\varepsilon) \le \sup_{U_{1:n}\le u\le\varepsilon} \frac{\mu Q(u)q(u)}{Q(L_F^{-1}(\tau_n(u)))} \sup_{0\le u\le\varepsilon} \frac{|\ell_n^*(u)|}{Q(u)q(u)}$$

with $\tau_n(u)$ as in (13.15). In force of Lemma 13.4 it is enough to show that the first supremum is stochastically bounded for each fixed ε . To do this, we must bound $\tau_n(u)$ or, what amounts to the same by (13.15), $C_n(u)$ from below for $u \ge U_{1:n}$ by a manageable quantity. Our starting point is again Wellner's (1978) Remark 1 (Lemma 2.7 here.) Let $\eta \in (0,1/2)$ be given. Then by Lemma 2 there exist two constants $K_1 = K_1(\eta)$ and $K_2 = K_2(\eta)$ such that $0 < K_1 < 1/2$, $K_1 < K_2$, and

(13.18) $$P\{K_1 \tfrac{i}{n} \le U_{i:n} \le K_2 \tfrac{i}{n}, \ 1 \le i \le n\} \ge 1-\eta$$

for each n. If

$$\omega \in \bigcap_{i=1}^{n} \{K_1 \tfrac{i}{n} \le U_{i:n} \le K_2 \tfrac{i}{n}\},$$

then on this ω we have for $i=1,\dots,n-1$ that

$$\frac{U_{i+1:n}}{U_{i:n}} \le \frac{K_2(i+1)}{K_1 i} \le 2\,\frac{K_2}{K_1},$$

and hence

$$\bigcap_{i=1}^{n} \{K_1 \tfrac{i}{n} \le U_{i:n} \le K_2 \tfrac{i}{n}\} \subset \bigcap_{i=1}^{n-1} \{U_{i+1:n} \le 2\,\frac{K_2}{K_1}\, U_i\}.$$

Also, by the strong law of large numbers there exists an n_1 such that

(13.19) $$P\{1/\bar{X}_n > 1/2\mu\} \ge 1-\eta, \quad n \ge n_1.$$

Now if

(13.20) $$\omega \in \bigcap_{i=1}^{n} \{K_1 \tfrac{i}{n} \le U_{i:n} \le K_2 \tfrac{i}{n}\} \cap \{\tfrac{1}{\bar{X}_n} > \tfrac{1}{2\mu}\},$$

and $u \in (U_{1:n}, 1]$, then we have

(13.21) $$U_{k:n} \leq u < U_{k+1:n}$$

for some $k = k(\omega,u) \geq 1$. Then on this ω,

$$\begin{aligned} C_n(u) &= \frac{1}{\bar{X}_n} \int_0^u Q(y)\,dE_n(y) \\ &\geq \frac{1}{2\mu} \int_0^u Q(y)\,dE_n(y) \\ &= \frac{1}{2\mu} \frac{1}{n} \sum_{i=1}^{k} Q(U_{i:n}) \\ &\geq \frac{1}{2\mu} \frac{1}{n} \sum_{i=1}^{k} Q(K_1 \tfrac{i}{n}) \\ &\geq \frac{1}{2\mu} \int_0^{k/n} Q(K_1 x)\,dx. \end{aligned}$$

Still on this ω,

$$\frac{K_1}{2K_2} U_{k+1:n} \leq U_{k:n} \leq K_2 \frac{k}{n},$$

and hence by (13.21)

$$\frac{k}{n} \geq \frac{K_1}{2K_2^2} u = \frac{u}{\lambda K_1}, \quad \lambda = \frac{2K_2^2}{K_1^2} > 1.$$

Thus, using also that $K_1 < 1/2$, we have

$$\begin{aligned} C_n(u) &\geq \frac{1}{2\mu} \int_0^{u/\lambda K_1} Q(K_1 x)\,dx \\ &= \frac{1}{2\mu K_1} \int_0^{u/\lambda} Q(y)\,dy \\ &> \frac{1}{\mu} \int_0^{u/\lambda} Q(y)\,dy \\ &= L_F(u/\lambda) \end{aligned}$$

on ω as in (13.20) and $u \geq U_{1:n}$. Taking therefore (13.18) and (13.19) into account, we have shown that for any $\eta \in (0,1/2)$ there exists a $\lambda = \lambda(\eta) > 1$ such that

$$\liminf_{n\to\infty} P\{C_n(u) \geq L_F(u/\lambda),\ U_{1:n} \leq u \leq 1\} \geq 1-2\eta.$$

Now the quantity

$$J(\varepsilon,\eta) = \sup_{0 \leq u \leq \varepsilon} \frac{Q(u)q(u)}{Q(u/\lambda(\eta))}$$

is finite, and hence, on considering also (13.15),

$$\limsup_{n\to\infty} P\{\sup_{U_{1:n}\le u\le\varepsilon} \frac{Q(u)q(u)}{Q(L_F^{-1}(\tau_n(u)))} > J(\varepsilon,\eta)\}$$

$$\le 2\eta + \limsup_{n\to\infty} P\{\sup_{U_{1:n}\le u\le\varepsilon} \frac{Q(u)q(u)}{Q(u/\lambda)} > J(\varepsilon,\eta)\}$$

$$= 2\eta .$$

This means that the first suprema in (13.17) are indeed bounded in probability, and hence

$$\lim_{\varepsilon\to 0}\limsup_{n\to\infty} P\{A_{4n}(\varepsilon) > \delta\} = 0$$

for any $\delta > 0$. Thus (13.13) is now proved.

The next step is to show that $\gamma_n(\cdot)$ and $\gamma_n(C_n^{-1}(L_F(\cdot)))$ are asymptotically the same. Indeed,

$$\sup_{0\le u\le 1}\left|\frac{\mu}{Q(u)}\Gamma_n^*(u) - \gamma_n(C_n^{-1}(L_F(u)))\right| \le$$

$$\le \sup_{0\le u\le 1}\left|\frac{\mu}{Q(C_n^{-1}(L_F(u)))}\Gamma_n^*(C_n^{-1}(L_F(u)))-\gamma_n(C_n^{-1}(L_F(u)))\right|$$

(13.22)

$$+ \sup_{0\le u\le 1}\left|\frac{\mu}{Q(u)}\Gamma_n^*(u) - \frac{\mu}{Q(C_n^{-1}(L_F(u)))}\Gamma_n^*(C_n^{-1}(L_F(u)))\right|$$

$$\le A_{2n} + \sup_{0\le u\le 1}\left|\frac{\mu}{Q(u)}\Gamma_n^*(u) - \frac{\mu}{Q(C_n^{-1}(L_F(u)))}\Gamma_n^*(C_n^{-1}(L_F(u)))\right| ,$$

and the case of A_{2n} was already settled in (13.13). Since (13.4) implies that

$$\sup_{0\le u\le 1}|C_n^{-1}(L_F(u))-u| \xrightarrow{a.s.} 0 ,$$

the second term also converges to zero in probability in view of the almost sure sample continuity of the process $\mu\Gamma^*(\cdot)/Q(\cdot)$, obtained by replacing B_n with B in the definition of Γ_n^* in (13.6), under the stated conditions.

Finally, we have to show that

(13.23)
$$A_{7n} = \sup_{0\le u\le 1}|\delta_n(u)| \xrightarrow{P} 0 ,$$

where

$$\delta_n(u) = n^{\frac{1}{2}}\{L_F^{-1}(C_n(C_n^{-1}(L_F(u))) - u\} .$$

First we observe that $\delta_n(\cdot)$ jumps at the points

$$u_k = L_F^{-1}\left(\frac{1}{n\bar{X}_n}\sum_{i=1}^{k} Q(U_{i:n})\right), \quad k=1,\ldots,n,$$

where $u_n = 1$, i.e., $\delta_n(\cdot)$ is constant on the intervals $[u_{k-1}, u_k)$, $k=1,\ldots,n$, where we define u_o to be zero. Now, if $u_{k-1} \le u < u_k$, then $C_n^{-1}(L_F(u)) = k/n$, $C_n(k/n) = L_F(u_k)$, that is, $\delta_n(u) = n^{\frac{1}{2}}(u_k - u)$. Whence

$$\begin{aligned} A_{7n} &\le \max_{1\le k\le n} \sup_{u_{k-1}\le u\le u_k} |\delta_n(u)| \\ &= \max_{1\le k\le n} n^{\frac{1}{2}}(u_k - u_{k-1}). \end{aligned}$$

On the other hand, going back to the process $\gamma_n(\cdot)$ in (13.12) we see that

$$\gamma_n(\tfrac{k}{n}) = n^{\frac{1}{2}}\{u_k - \tfrac{k}{n}\}, \quad \gamma_n(\tfrac{k}{n}-) = n^{\frac{1}{2}}\{u_{k-1} - \tfrac{k}{n}\}$$

for $k = 1,\ldots,n$. Thus

$$\begin{aligned} A_{7n} &\le \max_{1\le k\le n} \{\gamma_n(\tfrac{k}{n}) - \gamma_n(\tfrac{k}{n}-)\} \\ &\le \sup_{0\le u\le 1-\frac{1}{n}} \sup_{0\le u\le 1/n} |\gamma_n(u+y) - \gamma_n(u)| \\ &\le \sup_{0\le u\le 1-\frac{1}{n}} \sup_{0\le y\le 1/n} |\tilde{\Gamma}_n(u+y) - \tilde{\Gamma}_n(u)| \\ &\quad + 2 \sup_{0\le u\le 1} |\gamma_n(u) - \tilde{\Gamma}_n(u)|, \end{aligned} \tag{13.24}$$

where $\tilde{\Gamma}_n(u) = \mu\Gamma_n^*(u)/Q(u)$ are the processes in (13.13), and hence the second term in this upper bound converges to zero in probability. But since the processes $\tilde{\Gamma}_n$ are equally distributed as $\tilde{\Gamma}(\cdot) = \mu\Gamma^*(\cdot)/Q(\cdot)$, and

$$\sup_{0\le u\le 1-\frac{1}{n}} \sup_{0\le y\le\frac{1}{n}} |\tilde{\Gamma}(u+y) - \tilde{\Gamma}(y)| \xrightarrow{a.s.} 0$$

because $\tilde{\Gamma}$ is almost surely uniformly continuous on $[0,1]$, the first term also goes to zero in probability. Theorem 13.5 is now proved.

We defer the discussion of Theorem 13.5, that of its relationship to Goldie's (1977) corresponding result and the relationship of its proof to Vervaat's (1972) paper, until the next section.

Before stating our strong approximation result for the Goldie concentration process, we prove six lemmas to be used in the proof of that result. The first one is a very simple inequality for the Lorenz curve.

LEMMA 13.6. <u>For all</u> $\lambda \geq 1$,

$$\lambda L_F(u/\lambda) \leq L_F(u), \quad 0 \leq u \leq 1.$$

<u>Proof</u>. Since Q is nondecreasing,

$$\int_0^u Q(y)\,dy = \lambda \int_0^{u/\lambda} Q(\lambda x)\,dx$$

$$\geq \lambda \int_0^{u/\lambda} Q(x)\,dx ,$$

and the claim follows on dividing by the mean μ .

The next lemma is a James type result for the process ℓ_n^* in (13.5).

LEMMA 13.7. <u>If</u> $EX^2 < \infty$ <u>and</u> Q <u>is continuous on</u> $[0,1)$, <u>then</u>

$$\sup_{0 \leq u \leq 1} \left| \frac{n^{\frac{1}{2}}\{C_n(u) - L_F(u)\}}{Q(u)\,u^{\frac{1}{2}-\delta}} \right| \overset{a.s.}{=} O((\log\log n)^{\frac{1}{2}})$$

<u>for any</u> $\delta > 0$.

<u>Proof</u>. For $\ell_n^*(u)$ in the numerator we have

$$|\ell_n^*(u)| = |-\bar{X}_n^{-1} \int_0^u Q(y)\,d\alpha_n(y) + n^{\frac{1}{2}}(\bar{X}_n^{-1} - \mu^{-1}) \int_0^u Q(y)\,dy|$$

(13.25) $$\leq |\bar{X}_n^{-1}\alpha_n(u)Q(u)| + |\bar{X}_n^{-1} \int_0^u \alpha_n(y)\,dQ(y)|$$

$$+ \frac{uQ(u)}{\bar{X}_n \mu} \frac{1}{n^{\frac{1}{2}}} \left| \sum_{k=1}^{n} (X_i - \mu) \right| .$$

Dividing by $Q(u)u^{\frac{1}{2}-\delta}$ and taking sups termwise, the result for the first term follows by (2.8) and for the third one by the classical Hartman-Wintner log log law. The second term will be

(13.26) $$\frac{1}{\bar{X}_n} \sup_{0 \leq u \leq 1} \frac{1}{Q(u)} \left| \int_0^u \frac{\alpha_n(y)}{y^{\frac{1}{2}-\delta}}\,dQ(y) \right|$$

$$\leq \frac{1}{\bar{X}_n} \sup_{0 \leq u \leq 1} \frac{Q(u) - Q(0)}{Q(u)} \sup_{0 \leq u \leq 1} \frac{|\alpha_n(u)|}{u^{\frac{1}{2}-\delta}}$$

and hence the result for this second term again follows by (2.8).

The results for $\ell_n^*(n) = n^{\frac{1}{2}}\{C_n(u) - L_F(u)\}$ contained in the next lemma are analogous to Csáki's (1977) law of the iterated logarithm for the standardized empirical process, recorded here as Lemma 2.8, and will in fact be derived from Csáki's result. It will naturally appear in the proof of Theorem 13.12 below or, rather in that of Lemma 13.10, that we have to divide $\ell_n^*(u)$ by $(u(1-u))^{\frac{1}{2}}$, or by an even heavier function around zero. But this requires the extra conditions that

$$(13.27) \qquad 0 < \lim_{u \to 0} \frac{u^{\alpha}}{f(Q(u))} < \infty \quad \text{for some} \quad 0 \le \alpha < 1$$

and

$$(13.28) \qquad 0 < \lim_{u \to 1} \frac{(1-u)^{\beta}}{f(Q(u))} < \infty \quad \text{for some} \quad 0 \le \beta < 3/2 ,$$

where it is assumed, of course, that the indicated limits exist. It has already been discussed in Section 5 that the bound 3/2 is a natural one for β in condition (13.28). The latter discussion remains meaningful even if we now take the limit instead of simply taking supremum. On the other hand, assuming condition (13.27) with $\alpha \ge 1$ would at once imply that $Q(0) = -\infty$, and hence $t_F = -\infty$, although we are dealing with distributions having $t_F \ge 0$. So the restriction of α in $[0,1)$ is indeed necessary here in order to be consistent with the definition of the Lorenz curve and its inverse. Assuming conditions (13.27) and (13.28), thus strengthening the already familiar condition (5.2), reiterated as condition (10.9) when strongly approximating the empirical Lorenz process, results in the following inequalities:

$$(13.29) \qquad C_1 u^{1-\alpha} \le Q(u) - Q(0) \le C_2 u^{1-\alpha}, \quad 0 \le u \le 1/2,$$

and for $1/2 \le u \le 1$,

$$C_3 \le Q(u) \le C_4, \quad \text{if} \quad \beta < 1,$$

$$(13.30) \qquad C_5 \log \frac{1}{1-u} \le Q(u) \le C_6 \log \frac{1}{1-u} , \quad \text{if} \quad \beta = 1 ,$$

$$C_7 (1-u)^{1-\beta} \le Q(u) \le C_8 (1-u)^{1-\beta} , \quad \text{if} \quad \beta > 1 ,$$

for some positive constants $C_1, C_2, \ldots$. These inequalities complement those in (5.5) and (5.6) from below, and we know how important the latter were throughout in all strong approximation proofs. Since the quantile function Q figures in the denominator in the limit process of the concentration process, we need to know its rate of decrease to $Q(0)$ in the strong approximation results. Also, the nature of the inverse Lorenz process will require to know the rates of decrease (and increase) of the Lorenz curve L_F itself to zero (and to 1) when something more precise than weak convergence (as in Theorem 13.5) is to be proven. The latter however does not require extra conditions in addition to the ones above, for integrating the inequalities in (13.29) we obtain

$$(13.31) \qquad C_9 u \le L_F(u) \le C_{10} u , \quad 0 \le u \le 1/2, \quad \text{if} \quad Q(0) > 0,$$

and

$$C_{11}u^{2-\alpha} \le L_F(u) \le C_{12}u^{2-\alpha},\quad 0 \le u \le 1/2,\quad \text{if } Q(0) = 0, \tag{13.32}$$

while integrating the inequalities in (13.30) we get

$$\begin{aligned} &C_{13}(1-u) \le 1-L_F(u) \le C_{14}(1-u), && \text{if } \beta < 1,\\ &C_{15}(1-u)\log\frac{1}{1-u} \le 1-L_F(u) \le C_{16}(1-u)\log\frac{1}{1-u}, && \text{if } \beta = 1,\\ &C_{17}(1-u)^{2-\beta} \le 1-L_F(u) \le C_{18}(1-u)^{2-\beta}, && \text{if } \beta > 1, \end{aligned} \tag{13.33}$$

whenever $1/2 \le u \le 1$. All our C_k constants, here and later, are strictly positive.

LEMMA 13.8. _Let_ $\delta(n) = n^{-1}\log\log n$ _and assume that_ F _has a density function_ $f = F'$ _which is positive on the open support of_ F _and that conditions_ (13.27) _and_ (13.28) _are satisfied. Then, almost surely,_

$$\limsup_{n\to\infty} (\log\log n)^{-\frac{1}{2}} \sup_{\delta(n)\le u\le 1/2} \left|\frac{\ell_n^*(u)}{Q(u)u^{\frac{1}{2}}}\right| \le C_{19}$$

and if $\varepsilon(n) \ge \delta(n)$, _then_

$$\limsup_{n\to\infty} (\log\log n)^{-\frac{1}{2}} \sup_{1/2\le u\le 1-\varepsilon(n)} \left|\frac{\ell_n^*(u)}{(1-u)^{\frac{1}{2}}}\right| \le C_{20},\quad \underline{\text{if}}\ \ \beta < 1,$$

$$\limsup_{n\to\infty} (\varepsilon(n))^{\eta}(\log\log n)^{-\frac{1}{2}} \sup_{1/2\le u\le 1-\varepsilon(n)} \left|\frac{\ell_n^*(u)}{(1-u)^{\frac{1}{2}}}\right| \le C_{21},\quad \underline{\text{if}}\ \ \beta > 1,$$

where η _is an arbitrarily small positive number._

We call attention to the correspondence between Lemma 13.7 and the first statement of Lemma 13.8. By the extra conditions we were able to get free of $u^{-\delta}$ in the denominator, that is, we transferred a James type law into a Csáki type law. Nevertheless, it appears to be impossible to have such a law on $(0,1/2)$, and we have to remain above $\delta(n)$.

Proof. Using (13.25) above,

$$\begin{aligned} \sup_{\delta(n)\le u\le 1/2} \left|\frac{\ell_n^*(u)}{Q(u)u^{\frac{1}{2}}}\right| \le \frac{1}{\bar{X}_n} &\sup_{\delta(n)\le u\le 1/2} \frac{|\alpha_n(u)|}{u^{\frac{1}{2}}}\\ &+\frac{1}{\bar{X}_n} \sup_{\delta(n)\le u<1/2} \left|\frac{1}{Q(u)u^{\frac{1}{2}}} \int_0^{\delta(n)} \alpha_n(y)\,dQ(y)\right|\\ &+\frac{1}{\bar{X}_n} \sup_{\delta(n)<u\le 1/2} \left|\frac{1}{Q(u)u^{\frac{1}{2}}} \int_{\delta(n)}^{u} \alpha_n(y)\,dQ(y)\right| \end{aligned}$$

$$+ \frac{1}{\overline{X}_n} \frac{1}{n^{\frac{1}{2}}} \left| \sum_{k=1}^{n} (X_i - \mu) \right| \sup_{\delta(n) \le u \le 1/2} \frac{L_F(u)}{Q(u) u^{\frac{1}{2}}} .$$

The first term in this upper bound is

$$\text{(13.34)} \qquad \stackrel{a.s.}{=} O((\text{loglog } n)^{\frac{1}{2}})$$

by Lemma 2.8 and the strong law of large numbers, to be applied also in the subsequent terms without further mention. Using the inequalities in (13.29), (13.31) and (13.32) we see that the supremum in the fourth term of the said bound is bounded, and hence by the Hartmann-Wintner law of the iterated logarithm (condition (13.28) implies $EX^2 < \infty$) this fourth term is as in (13.34) again. The second term is not greater than

$$\frac{1}{Q(\delta(n))(\delta(n))^{\frac{1}{2}}} n^{\frac{1}{2}}\{E_n(\delta(n))+\delta(n)\}\{Q(\delta(n))-Q(0)\}$$

$$= \frac{Q(\delta(n))-Q(0)}{Q(\delta(n))} \left\{\frac{n^{\frac{1}{2}}[E_n(\delta(n))-\delta(n)]}{(\delta(n))^{\frac{1}{2}}} + 2(n\delta(n))^{\frac{1}{2}}\right\}$$

$$\stackrel{a.s.}{=} O((\text{loglog } n)^{\frac{1}{2}})$$

by Csáki's law in Lemma 2.8 again. The third term is not greater than

$$\frac{1}{\overline{X}_n} \sup_{\delta(n) \le u \le 1/2} \left| \frac{1}{Q(u)} \int_{\delta(n)}^{u} \frac{\alpha_n(y)}{y^{\frac{1}{2}}} dQ(y) \right|$$

$$\le \frac{1}{\overline{X}_n} \sup_{\delta(n) \le u \le 1/2} \frac{|\alpha_n(u)|}{u^{\frac{1}{2}}} \sup_{\delta(n) \le u \le 1/2} \frac{Q(u)-Q(\delta(n))}{Q(u)}$$

$$\stackrel{a.s.}{=} O((\text{loglog } n)^{\frac{1}{2}}),$$

again by Csáki's law. The first statement of the lemma is now proved.

The proof of the second three statements is based on the observation that since

$$\int_0^1 Q(y)dy = \mu \qquad \text{and} \qquad \int_0^1 Q(y)dE_n(y) = \overline{X}_n ,$$

we have

$$\ell_n^*(u) = n^{\frac{1}{2}}\left\{\frac{1}{\overline{X}_n} \int_0^u Q(y)dE_n(y) - \frac{1}{\mu} \int_0^u Q(y)dy\right\}$$

$$= n^{\frac{1}{2}}\left\{\frac{1}{\mu} \int_u^1 Q(y)dy - \frac{1}{\overline{X}_n} \int_u^1 Q(y)dE_n(y)\right\}$$

$$= \frac{1}{\overline{X}_n} n^{\frac{1}{2}}\left\{\int_u^1 Q(y)d(y-E_n(y)) + \left(\frac{1}{\mu} - \frac{1}{\overline{X}_n}\right) \int_u^1 Q(y)dy\right\}.$$

Hence, integrating by parts,

$$\sup_{1/2<u\leq 1-\varepsilon(n)} \left|\frac{\ell_n^*(u)}{(1-u)^{\frac{1}{2}}}\right| \leq \frac{1}{\bar{\bar{X}}_n} \sup_{1/2\leq u\leq 1-\varepsilon(n)} \frac{Q(u)\,|\alpha_n(u)|}{(1-u)^{\frac{1}{2}}}$$

$$+ \frac{1}{\bar{\bar{X}}_n} \sup_{1/2\leq u<1-\varepsilon(n)} \frac{1}{(1-u)^{\frac{1}{2}}} \int_u^1 |\alpha_n(y)|\,dQ(y)$$

$$+ \frac{1}{\bar{\bar{X}}_n\mu} \frac{1}{n^{\frac{1}{2}}} \left|\sum_{k=1}^{n} (X_k-\mu)\right| \sup_{1/2\leq u\leq 1-\varepsilon(n)} \frac{1}{(1-u)^{\frac{1}{2}}} \int_u^1 Q(y)\,dy$$

$$= A_{8n} + A_{9n} + A_{10n}.$$

By Lemma 2.8 and the upper bounds in (13.30),

$$A_{8n} \overset{a.s.}{=} O((\log\log n)^{\frac{1}{2}})Q(1-\varepsilon(n))$$

$$= \begin{cases} O((\log\log n)^{\frac{1}{2}}) & , \text{ if } \beta < 1, \\ O((\log\log n)^{\frac{1}{2}}\log(\varepsilon(n))^{-1}), & \text{ if } \beta = 1, \\ O((\log\log n)^{\frac{1}{2}}(\varepsilon(n))^{1-\beta}) & , \text{ if } \beta > 1. \end{cases}$$

Since

$$\frac{1}{\mu}\int_u^1 Q(y)\,dy = 1 - L_F(u),$$

the Hartmann-Wintner law and the upper bounds in (13.30) give that

$$A_{10n} \overset{a.s.}{=} O((\log\log n)^{\frac{1}{2}}) \sup_{1/2\leq u\leq 1-\varepsilon(n)} \frac{1-L_F(u)}{(1-u)^{\frac{1}{2}}}$$

$$= O((\log\log n)^{\frac{1}{2}})$$

for any $\beta \in [0,3/2)$.

In the estimation of A_{9n} we separate the cases of different β from the beginning. If $\beta < 1$, then

$$A_{9n} \leq \frac{1}{\bar{\bar{X}}_n} \sup_{1/2\leq u\leq 1-\delta(n)} \frac{1}{(1-u)^{\frac{1}{2}}} \int_u^{1-\delta(n)} |\alpha_n(y)|\,dQ(y)$$

$$+ \frac{1}{\bar{\bar{X}}_n} \frac{1}{(\delta(n))^{\frac{1}{2}}} \int_{1-\delta(n)}^1 |\alpha_n(y)|\,dQ(y)$$

$$\leq \frac{1}{\bar{\bar{X}}_n} (Q(1)-Q(\tfrac{1}{2}))\Big\{ \sup_{1/2\leq u\leq 1-\delta(n)} \frac{|\alpha_n(u)|}{(1-u)^{\frac{1}{2}}}$$

$$+ \frac{1}{(1-(1-\delta(n)))^{\frac{1}{2}}} \sup_{1-\delta(n)\leq u\leq 1} |\alpha_n(u)|\Big\}.$$

An elementary computation now shows that the second term in the bracket { } is not greater than

$$\frac{n^{\frac{1}{2}}\{1-E_n(1-\delta(n)) - (1-(1-\delta(n)))\}}{(1 - (1-\delta(n)))^{\frac{1}{2}}} + 2(n\delta(n))^{\frac{1}{2}},$$

and therefore by Lemma 2.8 again

$$A_{9n} \stackrel{a.s.}{=} O((\log\log n)^{\frac{1}{2}}), \quad \text{if} \quad \beta < 1.$$

In the same way as above, but keeping $\varepsilon(n)$, if $\beta \geq 1$,

$$A_{9n} \leq \frac{1}{\overline{X}_n} \sup_{1/2 \leq u \leq 1-\varepsilon(n)} \frac{1}{(1-u)^{\frac{1}{2}}} \int_u^{1-\varepsilon(n)} |\alpha_n(y)|\,dQ(y)$$

$$+ \frac{1}{\overline{X}_n} \frac{1}{(\varepsilon(n))^{\frac{1}{2}}} \int_{1-\varepsilon(n)}^{1} |\alpha_n(y)|\,dQ(y)$$

$$= A_{11n} + A_{12n}.$$

Since $\varepsilon(n) \geq \delta(n)$, Lemma 2.8 again implies that

$$A_{11n} \stackrel{a.s.}{=} O((\log\log n)^{\frac{1}{2}})Q(1-\varepsilon(n))$$

$$= \begin{cases} O((\log\log n)^{\frac{1}{2}} \log \frac{1}{\varepsilon(n)})), & \text{if} \quad \beta = 1, \\ O((\log\log n)^{\frac{1}{2}}(\varepsilon(n))^{1-\beta}), & \text{if} \quad \beta > 1. \end{cases}$$

Using now condition (13.28) directly,

$$A_{12n} \leq \frac{C_{23}}{\overline{X}_n} \frac{1}{(\varepsilon(n))^{\frac{1}{2}}} \int_{1-\varepsilon(n)}^{1} |\alpha_n(y)|(1-y)^{-\beta}\,dy$$

$$\leq \frac{C_{23}}{\overline{X}_n} \sup_{1-\varepsilon(n) \leq y \leq 1} \frac{|\alpha_n(y)|}{(1-y)^{\frac{1}{2}-\delta}} \frac{1}{(\varepsilon(n))^{\frac{1}{2}}} \int_{1-\varepsilon(n)}^{1} (1-y)^{\frac{1}{2}-\beta-\delta}\,dy$$

$$\stackrel{a.s.}{=} O((\log\log n)^{\frac{1}{2}}(\varepsilon(n))^{1-\beta-\delta})$$

for any small $\delta > 0$ by James' law in (2.8). Collecting now all the bounds, the second three statements of the lemma are also proved.

It is clear from the proof of Theorem 13.5 that the basic ingredient of the concentration process is the process

$$\gamma_n(u) = n^{\frac{1}{2}}\{L_F^{-1}(C_n(u))-u\}, \quad 0 \leq u \leq 1,$$

of (13.12). When strengthening the proof of Theorem 13.5 for the sake of strong approximation of $c_n(\cdot)$, we must approximate γ_n. We separate this part as the main body in the proof of Theorem 13.12 and formulate it in Lemma 13.10 below, which is a strong form of (13.13). This result in turn requires an already almost established approximation for ℓ_n^* of (13.5) (and the preceding lemmas). As a strong form of Lemma 13.3 we have the following result.

LEMMA 13.9. Under the conditions of Theorem 10.3 (or 11.3)

$$\Delta_n^{(10)} = \sup_{0\le u\le 1} |\ell_n^*(u)-\Gamma_n^*(u)| \overset{a.s.}{=} O(n^{-\lambda}),$$

where $\lambda < \min(1/2, \frac{3}{2} - \beta)$.

Proof. Using the bound for $\Delta_n^{(10)}$ right below (13.7), we have

$$\Delta_n^{(10)} \le \frac{2}{\bar{X}_n}(\Delta_n^{(2)} + \Delta_{1n}) + |\frac{1}{\bar{X}_n} - \frac{1}{\mu}| \sup_{0\le u\le 1} |\Gamma_n^*(u)|,$$

where $\Delta_n^{(2)}$ and Δ_{1n} are exactly as in (10.5). Hence, by (10.11), (10.15) and the Hartman-Wintner loglog law,

$$\Delta_n^{(10)} \overset{a.s.}{=} O(n^{-\lambda}+(n^{-1}\text{loglog } n)^{\frac{1}{2}} \sup_{0\le u\le 1} |\Gamma_n^*(u)|).$$

On integrating by parts, multiplying and dividing by $(1-u)^{\frac{1}{2}-\delta}$, where $\delta > 0$ is arbitrarily small, and using (2.9) we get

$$\sup_{0\le u\le 1} |\Gamma_n^*(u)| \le \sup_{0\le u\le 1} \frac{|B_n(u)|}{(1-u)^{\frac{1}{2}-\delta}}\{\frac{1}{\mu}\int_0^1 (1-y)^{-\frac{1}{2}+\delta}dQ(y)$$

$$\text{(13.35)} \qquad + \sup_{0\le u\le 1} |(1-u)^{\frac{1}{2}-\delta} Q(u)|\}$$

$$\overset{a.s.}{=} O((\text{loglog } n)^{\frac{1}{2}})$$

in view of the fact that the terms in { } are bounded as a consequence of the inequalities in (5.6).

LEMMA 13.10. Suppose that F has a density function $f = F'$, positive on the open support of F, and conditions (13.27) and (13.28) are satisfied. Then

$$A_{2n} = \sup_{0\le u\le 1} |\frac{\mu}{Q(u)} \Gamma^*(u)-\gamma_n(u)| \overset{a.s.}{=} O(n^{-\rho}),$$

where $\rho = \min(\rho_1,\rho_2)$ with

$$\rho_1 = \begin{cases} \lambda, & \text{if } Q(0) > 0, \\ \frac{\lambda}{3-2\alpha}, & \text{if } Q(0) = 0, \end{cases} \quad \text{and} \quad \rho_2 < \begin{cases} \frac{1}{2(2\beta+1)}, & \text{if } \beta < 1, \\ \frac{3-2\beta}{4(\beta^2-1)+6}, & \text{if } \beta \ge 1, \end{cases}$$

where $\lambda < \min(\frac{1}{2}, \frac{3}{2} - \beta)$ is of Lemma 13.9.

Proof. Using more cutting points than in (13.14), we get

$$A_{2n} \le \sup_{0\le u\le U_{1:n}} |\gamma_n(u)| + \sup_{U_{1:n}\le u\le \delta_1(n)} |\gamma_n(u)|$$

$$+ \sup_{0\le u\le \varepsilon_{1n}} \left|\frac{\mu}{Q(u)} \Gamma_n^*(u)\right| + \sup_{\delta_1(n)\le u\le \delta_{1n}} |\gamma_n(u)|$$

$$+ \sup_{\varepsilon_{1n}\le u\le 1/2} \left|\frac{\mu}{Q(u)} \Gamma_n^*(u) - \gamma_n(u)\right|$$

$$+ \sup_{1-\varepsilon_{2n}\le u\le 1} \left|\frac{\mu}{Q(u)} \Gamma_n^*(u)\right|$$

$$+ \sup_{1-\varepsilon_{2n}\le u\le 1} |\gamma_n(u)|$$

$$= A_{3n} + A_{4n} + A_{5n} + A_{13n} + \ldots + A_{17n},$$

where the cutting sequences will be defined below. As we know,

(13.36) $$A_{3n} = n^{\frac{1}{2}} U_{1:n} \stackrel{a.s.}{=} O(n^{-\frac{1}{2}} \log n),$$

on applying Lemma 2.10.

Letting

$$\delta_1(n) = \begin{cases} \left(\frac{3C_{19}}{C_9} Q(\tfrac{1}{2})\right)^2 \frac{\log\log n}{n}, & \text{if } Q(0) > 0, \\ \left(\frac{3C_{19}}{C_{11}} C_2\right)^2 \frac{\log\log n}{n}, & \text{if } Q(0) = 0, \end{cases}$$

where C_2, C_9, C_{11} and C_{19} are as in (13.29), (13.31), (13.32) and Lemma 13.8, respectively, we have, just as in (13.17), that

$$A_{4n} \le \sup_{U_{1:n}\le u\le \delta_1(n)} \frac{\mu Q(u) u^{\frac{1}{2}-\delta}}{Q(L_F^{-1}(\tau_n(u)))} \sup_{0\le u\le \delta_1(n)} \frac{|\ell_n^*(u)|}{Q(u) u^{\frac{1}{2}-\delta}}$$

(13.37) $$\stackrel{a.s.}{=} O((\log\log n)^{\frac{1}{2}}) \frac{\mu Q(\delta_1(n))(\delta_1(n))^{\frac{1}{2}-\delta}}{Q(L_F^{-1}(\tau_n(U_{1:n})))}$$

by Lemma 13.7. Of course, $\tau_n(\cdot)$ here is as in (13.15), and we must find a lower bound for $C_n(U_{1:n})$. Making use of Lemmas 2.10 and 13.6,

$$C_n(U_{1:n}) = \frac{1}{\bar{X}_n} \frac{1}{n} Q(U_{1:n})$$

$$= \frac{\mu}{\bar{X}_n} \frac{1}{nU_{1:n}} \frac{1}{\mu} U_{1:n} Q(U_{1:n})$$

$$\geq \frac{\mu}{\bar{X}_n} \frac{1}{nU_{1:n}} \frac{1}{\mu} \int_0^{U_{1:n}} Q(y)\,dy$$

$$\geq (\log n)^{-2} L_F(U_{1:n}) \quad \text{(almost surely)}$$

$$\geq L_F((\log n)^{-2} U_{1:n}).$$

Whence, and from (13.15),

$$Q(L_F^{-1}(\tau_n(U_{1:n}))) \geq Q((\log n)^{-2} U_{1:n}), \quad \text{a.s.},$$

that is, if $Q(0) = 0$ then

$$\frac{Q(\delta_1(n))}{Q(L_F^{-1}(\tau_n(U_{1:n})))} \leq \frac{C_2}{C_1} \left(\frac{\delta_1(n)}{(\log n)^{-2} U_{1:n}}\right)^{1-\alpha}$$

$$= O((\log n)^{4(1-\alpha)} (\log\log n)^{1-\alpha})$$

almost surely, by (13.29) and Lemma 2.10. If $Q(0) > 0$, then the latter ratio of the Q's is obviously bounded. Returning then to (13.37), we obtained that

(13.38) $$A_{4n} \overset{\text{a.s.}}{=} O(n^{-\frac{1}{2}+\delta})$$

for any small $\delta > 0$.

The already familiar term A_{5n} is easy. We introduce ε_{1n} as $\varepsilon_{1n} = n^{-\varepsilon_1}$ where

$$0 < \varepsilon_1 < \begin{cases} 2\lambda, & \text{if } Q(0) > 0, \\ \frac{2\lambda}{3-2\alpha}, & \text{if } Q(0) = 0, \end{cases}$$

with λ as in Lemma 13.9. Then, after integrating by parts, we see that

$$A_{5n} \leq \sup_{0 \leq u \leq \varepsilon_{1n}} \frac{L_F(u)}{Q(u)} \left|\int_0^1 B_n(y)\,dQ(y)\right|$$

$$+ \sup_{0 \leq u \leq \varepsilon_{1n}} |B_n(u)| \sup_{0 \leq u \leq \varepsilon_{1n}} \left(\frac{Q(u)-Q(0)}{Q(u)} + 1\right).$$

Because of the inequalities in (13.29), (13.31) and (13.32),

$$\frac{L_F(u)}{Q(u)} \leq C_{24} u, \quad 0 \leq u \leq 1/2,$$

in either of the cases $Q(0) = 0$ or $Q(0) > 0$, and hence by (13.35) and (2.11) we get

(13.39) $$A_{5n} \overset{a.s.}{=} O((\varepsilon_{1n} \log n)^{\frac{1}{2}}).$$

Now, in order to be able to handle A_{13n} and A_{14n}, we first show that

(13.40) $$\limsup_{n\to\infty} \sup_{\delta_1(n)\le u\le 1/2} \frac{L_F(u)}{C_n(u)} \le 2 \quad \text{a.s.},$$

which statement corresponds roughly to Lemma 2.9. We may and do assume that C_{19} of Lemma 13.8 is chosen so large that $\delta_1(n) \ge \delta(n) = n^{-1}\log\log n$. Lemma 13.8 then implies that

$$|L_F(u)-C_n(u)| \le \tfrac{3}{2} C_{19}Q(u)(un^{-1}\log\log n)^{\frac{1}{2}}, \quad \delta_1(n) \le u \le 1/2,$$

which, in turn, implies that

(13.41a) $$\frac{L_F(u)}{C_n(u)} \le 1 + \frac{\frac{3}{2} C_{19}Q(u)(un^{-1}\log\log n)^{\frac{1}{2}}}{L_F(u) - \frac{3}{2} C_{19}Q(u)(un^{-1}\log\log n)^{\frac{1}{2}}}$$

for $\delta_1(n) \le u \le 1/2$ almost surely if $n \ge n_o(\omega)$.

First consider the case when $Q(0) > 0$. Then (13.31) and (13.41a) give

$$\frac{L_F(u)}{C_n(u)} \le 1 + \frac{\frac{3}{2} C_{19}Q(\frac{1}{2})(un^{-1}\log\log n)^{\frac{1}{2}}}{C_9 u - \frac{3}{2} C_{19}Q(\frac{1}{2})(un^{-1}\log\log n)^{\frac{1}{2}}}$$

$$= 1 + \frac{\frac{3}{2} C_{19}Q(\frac{1}{2})(u^{-1}n^{-1}\log\log n)^{\frac{1}{2}}}{C_9 - \frac{3}{2} C_{19}Q(\frac{1}{2})(u^{-1}n^{-1}\log\log n)^{\frac{1}{2}}}$$

$$\le 2, \quad \delta_1(n) \le u \le 1/2,$$

almost surely if $n \ge n_o(\omega)$ upon replacing u by its smallest value $\delta_1(n)$ and using the definition of $\delta_1(n)$.

When $Q(0) = 0$, then again by (13.41a), and by (13.29) and (13.32) we arrive at

$$\frac{L_F(u)}{C_n(u)} \le 1 + \frac{\frac{3}{2} C_{19}C_2u^{1-\alpha}(un^{-1}\log\log n)^{\frac{1}{2}}}{C_{11}u^{2-\alpha} - \frac{3}{2} C_{19}C_2u^{1-\alpha}(un^{-1}\log\log n)^{\frac{1}{2}}}$$

$$= 1 + \frac{\frac{3}{2} C_{19}C_2(u^{-1}n^{-1}\log\log n)^{\frac{1}{2}}}{C_{11} - \frac{3}{2} C_{19}C_2(un^{-1}\log\log n)^{\frac{1}{2}}}$$

$$\le 2, \quad \delta_1(n) \le u \le 1/2,$$

almost surely if $n \ge n_o(\omega)$. Thus (13.40) is indeed true.

We now turn to A_{13n}. By the one-term Taylor formula as in

(13.37),

$$A_{13n} \leq \sup_{\delta_1(n) \leq u \leq \varepsilon_{1n}} \frac{\mu Q(u) u^{\frac{1}{2}-\delta}}{Q(L_F^{-1}(\tau_n(u)))} \sup_{\delta_1(n) \leq u \leq \varepsilon_{1n}} \frac{|\ell_n^*(u)|}{Q(u) u^{\frac{1}{2}-\delta}},$$

where by (13.15), (13.40) and Lemma 13.6

$$\text{(13.41b)} \qquad \tau_n(u) \geq \frac{1}{3} L_F(u) \geq L_F(u/3), \quad \delta_1(n) \leq u \leq 1/2.$$

Thus

$$\sup_{\delta_1(n) \leq u \leq 1/2} \frac{Q(u)}{Q(L_F^{-1}(\tau_n(u)))} < \infty \quad \text{a.s.},$$

and then

$$\text{(13.42)} \qquad A_{13n} \overset{\text{a.s.}}{=} O((\varepsilon_{1n})^{\frac{1}{2}-\delta}(\log\log n)^{\frac{1}{2}}),$$

for any small $\delta > 0$, follows from Lemma 13.7.

Next we consider A_{14n}. A two-term Taylor formula leads to

$$\begin{aligned} A_{14n} &\leq \sup_{\varepsilon_{1n} \leq u \leq 1/2} \frac{\mu}{Q(u)} |\ell_n^*(u) - \Gamma_n^*(u)| \\ \text{(13.43)} \qquad &+ \sup_{\varepsilon_{1n} \leq u \leq 1/2} \frac{\mu n^{-\frac{1}{2}} |\ell_n^*(u)|^2}{Q^3(L_F^{-1}(\tau_n(u))) f(Q(L_F^{-1}(\tau_n(u))))} \\ &= A_{18n} + A_{19n}, \end{aligned}$$

where

$$A_{18n} \overset{\text{a.s.}}{=} \begin{cases} O(n^{-\lambda}) & , \quad Q(0) > 0, \\ O((\varepsilon_{1n})^{\alpha-1} n^{-\lambda}), & \text{if} \quad Q(0) = 0, \end{cases}$$

as a consequence of Lemma 13.9 and (13.29). An obvious manipulation yields

$$\begin{aligned} A_{19n} &\leq \mu n^{-\frac{1}{2}} \sup_{\varepsilon_{1n} \leq u \leq 1/2} \left(\frac{Q(u)}{Q(L_F^{-1}(\tau_n(u)))} \right)^2 \sup_{\varepsilon_{1n} \leq u \leq 1/2} \frac{(L_F^{-1}(\tau_n(u)))^{\alpha}}{f(Q(L_F^{-1}(\tau_n(u))))} \\ &\times \sup_{\varepsilon_{1n} \leq u \leq 1/2} \frac{\mu}{L_F^{-1}(\tau_n(u))} \sup_{\varepsilon_{1n} \leq u \leq 1/2} \frac{(L_F^{-1}(\tau_n(u)))^{1-\alpha}}{Q(L_F^{-1}(\tau_n(u)))} \\ \text{(13.44)} \qquad &\times \sup_{0 \leq u \leq 1/2} \left(\frac{\ell_n^*(u)}{Q(u) u^{\frac{1}{2}-\delta}} \right)^2 u^{-2\delta} \\ &\overset{\text{a.s.}}{=} O((\varepsilon_{1n})^{-2\delta} n^{-\frac{1}{2}} \log\log n), \end{aligned}$$

by Lemma 13.7 for the first, while the fourth suprema are bounded by (13.41b) and (13.29), and the second and third suprema are bounded by (13.27) and (13.41b) respectively. The last two order relations and the definition of λ imply that

$$(13.45) \qquad A_{14n} \overset{a.s.}{=} \begin{cases} O(n^{-\lambda}) & , \text{ if } Q(0) > 0, \\ O((\varepsilon_{1n})^{\alpha-1} n^{-\lambda}), & \text{if } Q(0) = 0, \end{cases}$$

for $\delta > 0$ was arbitrarily small.

It is worthwhile to collect the so far produced rates now, since they depend on α and λ only. We have

$$(13.46) \qquad \begin{aligned} A_{2n} \le & \sup_{0 \le u < 1/2} \left| \frac{\mu}{Q(u)} \Gamma_n^*(u) - \gamma_n(u) \right| \\ & + A_{15n} + A_{16n} + A_{17n}, \end{aligned}$$

where (13.36), (13.38), (13.39), (13.42) and (13.45) together give

$$(13.47) \qquad \sup_{0 \le u \le 1/2} \left| \frac{\mu}{Q(u)} \Gamma_n^*(u) - \gamma_n(u) \right| \overset{a.s.}{=} O(n^{-\rho_1}),$$

and ρ_1 is now as in the statement of the lemma.

In order to be able now to estimate the remaining terms, all corresponding to the upper half of the problem, that is, to the interval $[1/2,1]$, we have to prove a corresponding analogue of (13.40).

Towards this end, set

$$\delta_2(n) = \begin{cases} \left(\dfrac{3C_4C_{20}}{C_{14}}\right)^2 n^{-1} \log\log n, & \text{if } \beta < 1, \\ n^{-1+\delta} & , \text{ if } \beta \ge 1, \end{cases}$$

where $\delta \in (0,1)$ is arbitrarily close to zero, and C_{20} of Lemma 13.8 is taken so large that $\delta_2(n) \ge \delta(n) = n^{-1} \log\log n$ also in the case of $\beta < 1$. Then

$$(13.48) \qquad \limsup_{n \to \infty} \sup_{1/2 \le u \le 1-\delta_2(n)} \frac{1-L_F(u)}{1-C_n(u)} \le 2 \quad \text{a.s.}$$

The proof of this result is completely analogous to that of 13.40: the inequalities in (13.30) and (13.33) together with the second three statements of Lemma 13.8 lead to

$$\frac{1-L_F(u)}{1-C_n(u)} \le 2, \quad \frac{1}{2} \le u \le 1-\delta_2(n),$$

almost surely for large enough (random) n. This implies (13.48), but using (13.33), it also implies that

$$(13.49)\qquad \frac{1-u}{1-L_F^{-1}(C_n(u))} \le \begin{cases} C_{25} & , \quad \beta < 1, \\ C_{26}\log\frac{1}{1-u} & , \quad \beta = 1, \\ C_{27}(1-u)^{1-\beta} & , \quad \beta > 1, \end{cases}$$

for $1/2 \le u \le 1-\delta_2(n)$ almost surely for all sufficiently large n beginning from some random threshold.

Now we start estimating A_{15n}. Similarly as in (13.43) and (13.44),

$$A_{15n} \le O(n^{-\lambda}) + \mu n^{-\frac{1}{2}} \sup_{1/2\le u\le 1-\varepsilon_{2n}} \frac{1}{(Q(L_F^{-1}(\tau_n(u))))^3}$$

$$\times \sup_{1/2\le u\le 1-\varepsilon_{2n}} \frac{(1-L_F^{-1}(\tau_n(u)))^{\beta}}{f(Q(L_F^{-1}(\tau_n(u))))}$$

$$\times \sup_{1/2\le u\le 1-\varepsilon_{2n}} \left(\frac{1-u}{1-L_F^{-1}(\tau_n(u))}\right)^{\beta}$$

$$\times \sup_{1/2\le u\le 1-\varepsilon_{2n}} (\ell_n^*(u))^2(1-u)^{-\beta}$$

almost surely with $\tau_n(u)$ as in (13.15), where $\varepsilon_{2n} = n^{-\varepsilon_2}$ and

$$\varepsilon_2 < \begin{cases} \frac{1}{2\beta+1}, & \text{if } \beta \le 1. \\ \frac{1}{2\beta^2-2\beta+3}, & \text{if } \beta > 1. \end{cases}$$

The first supremum factor is trivially bounded, while the second one is bounded by conditions (13.28). Since $\varepsilon_{2n} \ge \delta_2(n)$, the third supremum is

$$\overset{a.s.}{=} \begin{cases} O(1) & , \quad \beta < 1, \\ O((\log \frac{1}{\varepsilon_{2n}})^{\beta}) & , \quad \beta = 1, \\ O(\varepsilon_{2n}^{\beta(1-\beta)}) & , \quad \beta > 1, \end{cases}$$

by (13.49). Using now Lemma 13.9 and (13.35) jointly in order to obtain the rate of $|\ell_n^*(\cdot)|^2$ in the fourth supremum, we get that this supremum is

$$\overset{a.s.}{=} O(\varepsilon_{2n}^{-\beta}\log\log n).$$

Altogether,

$$(13.50)\qquad A_{15n} \overset{a.s.}{=} \begin{cases} O(n^{-\frac12}\varepsilon_{2n}^{-\beta} \text{ loglog } n), & \beta < 1, \\ O(n^{-\frac12}\varepsilon_{2n}^{-\beta}(\log \varepsilon_{2n}^{-1})^{\beta} \text{loglog } n), & \beta = 1, \\ O(n^{-\frac12}\varepsilon_{2n}^{-\beta^2} \text{ loglog } n), & \beta > 1, \end{cases}$$

for simple computations show that these rates are bigger than $n^{-\lambda}$, $\lambda < \min(\frac{1}{2}, \frac{3}{2} - \beta)$.

Next we consider the Gaussian cut A_{16n}. Just as $\ell_n^*(u)$ could be written in the proof of Lemma 13.8 as integrals around 1, the key is a corresponding form for Γ_n^* :

$$(13.51)\qquad \frac{\mu}{Q(u)} \Gamma_n^*(u) = \frac{1}{Q(u)} \{(L_F(u)-1) \int_0^1 Q(y)dB_n(y) + \int_u^1 Q(y)dB_n(y)\}$$

$$= \frac{1}{Q(u)} \{(1-L_F(u)) \int_0^1 \frac{B_n(y)}{f(Q(y))} dy - Q(u)B_n(u) - \int_u^1 \frac{B_n(y)}{f(Q(y))} dy\}.$$

Routine computation based on (13.33), (2.9), (2.11) and (13.28) leads to the rate of convergence

$$(13.52)\qquad A_{16n} \overset{a.s.}{=} \begin{cases} O((\varepsilon_{2n} \text{ loglog } n)^{\frac12}), & \beta < 1, \\ O(\varepsilon_{2n}^{\frac32-\beta-\delta} (\text{loglog } n)^{\frac12}), & \beta \geq 1. \end{cases}$$

The rate of A_{17n} is exactly the same if we use the just mentioned form of ℓ_n^* and (2.8) instead of (2.9). Hence the bounds in (13.50) and (13.52) and the definition of ε_{2n} give (see (13.46)), with ρ_2 already as in the formulation of the lemma,

$$\sup_{1/2 \leq u \leq 1} |\frac{\mu}{Q(u)} \Gamma_n^*(u) - \gamma_n(u)| \overset{a.s.}{=} O(n^{-\rho_2})$$

after some elementary computation and in view of the fact that $\delta > 0$ is arbitrarily small. This and (13.47) prove Lemma 13.10.

We recall that in the last stage of the proof of the weak approximation in Theorem 13.5 we needed to know that a uniform modulus of continuity of the process

$$\tilde{\Gamma}_n(u) = \frac{\mu}{Q(u)} \Gamma_n^*(u) , \quad 0 \leq u \leq 1,$$

converged to zero. Accordingly, we now need to have a rate result for such a modulus.

LEMMA 13.11. Suppose that conditions (13.27) and (13.28) hold. Then there exists a positive constant $C_{28} < \infty$ depending only on α and β such that

$$\limsup_{n\to\infty} (\log n)^{-\frac{1}{2}} \sup_{0\le u\le 1-h} \sup_{0\le y\le h} |\tilde{\Gamma}_n(u+y)-\tilde{\Gamma}_n(u)| \le C_{28}\, h^{\frac{1}{3}-\delta}$$

almost surely for all $h \in (0, \sqrt{2}/4)$ and $\delta > 0$.

Proof. We cut the problem into two pieces and investigate the two quantities

$$\Delta_1(n,h) = \sup_{0\le u\le 1/2} \sup_{0\le y\le h} |\tilde{\Gamma}_n(u+y) - \tilde{\Gamma}_n(u)|$$

and

$$\Delta_2(n,h) = \sup_{1/2\le u\le 1-h} \sup_{0\le y\le h} |\tilde{\Gamma}_n(u+y) - \tilde{\Gamma}_n(u)|$$

separately.

$$\Delta_1(n,h) \le 2 \sup_{0\le u\le 2h^{2/3}} |\tilde{\Gamma}_n(u)|$$

$$+ \sup_{h^{2/3}\le u\le 1/2} \sup_{0\le y\le h} |\tilde{\Gamma}_n(u+y) - \tilde{\Gamma}_n(u)|$$

$$= \Delta_1^{(1)}(n,h) + \Delta_1^{(2)}(n,h),$$

and via integrating by parts,

$$\Delta_1^{(1)}(n,h) \le 2 \sup_{0\le u\le 2h^{2/3}} \frac{L_F(u)}{Q(u)} \left|\int_0^1 \frac{B_n(y)}{f(Q(y))}\, dy\right|$$

$$+ 2 \sup_{0\le u\le 2h^{2/3}} |B_n(u)|\left(1 + \frac{Q(u)-Q(0)}{Q(u)}\right)$$

$$\underset{=}{\text{a.s.}}\ h^{2/3} O((\log\log n)^{\frac{1}{2}})$$

$$+ h^{1/3} O((\log n)^{\frac{1}{2}})$$

by (13.29), (13.31), (13.32), (2.9) and (13.27) for the first term, and by (2.11) for the second term. On the other hand, again integration by parts and straightforward manipulations yield

$$\Delta_1^{(2)}(n,h) \le \left|\int_0^1 B_n(y)\, dQ(y)\right| \Big\{ \sup_{h^{2/3}\le u\le 1/2} \sup_{0\le y\le h} \frac{1}{Q(u)}(L_F(u+y) - L_F(u))$$

$$+ \sup_{h^{2/3}\le u\le 1/2} \sup_{0\le y\le h} L_F(u+y)\left(\frac{1}{Q(u)} - \frac{1}{Q(u+y)}\right)\Big\}$$

$$+ \sup_{h^{2/3} \le u \le 1/2} \sup_{0 \le y \le h} |B_n(u) - B_n(u+y)|$$

$$+ \sup_{h^{2/3} \le u \le 1/2} \sup_{0 \le y \le h} \frac{1}{Q(u)} \left| \int_u^{u+y} B_n(t)\,dQ(t) \right|$$

$$+ \sup_{h^{2/3} \le u \le 1/2} \sup_{0 \le y \le h} \left(\frac{1}{Q(u)} - \frac{1}{Q(u+y)}\right) \left| \int_0^{u+y} B_n(t)\,dQ(t) \right|$$

$$= \Delta_1^{(3)}(n,h) + \ldots + \Delta_1^{(6)}(n,h).$$

Since, on utilizing (13.29),

$$\sup_{0 \le y \le h} \frac{1}{Q(u)} (L_F(u+y) - L_F(u)) = \frac{1}{\mu} \sup_{0 \le y \le h} \frac{1}{Q(u)} \int_u^{u+y} Q(t)\,dt$$

$$\le \frac{h}{\mu} \sup_{0 \le y \le h} \frac{Q(u+y)}{Q(u)}$$

$$\le C_{29} h$$

in view of the fact that here $u \le h$, and since

$$\sup_{0 \le y \le h} L_F(u+y) \left(\frac{1}{Q(u)} - \frac{1}{Q(u+y)}\right)$$

$$\le h \sup_{0 \le y \le h} \frac{(\xi(y))^{\alpha}}{f(Q(\xi(y)))} \frac{1}{Q(u)Q(u+y)} \left(\frac{u}{\xi(y)}\right)^{\alpha} L_F(u+y) u^{-\alpha}$$

$$\le C_{30} h$$

after a one-term Taylor, where $u \le \xi(u) \le u+h$, using (13.29), (13.31) and (13.32), we get

$$\Delta_1^{(3)}(u,h) \overset{a.s.}{=} hO((\log\log n)^{1/2}).$$

Lemma 5.4 states that $\Delta_1^{(4)}(n,h) \overset{a.s.}{=} h^{1/2} O((\log n)^{1/2})$. Taking out $\sup |B_n(t)|$ from the integral and applying again the Taylor formula for $Q(u+y) - Q(u)$, we obtain

$$\Delta_1^{(5)}(n,h) \le \sup_{0 \le t \le 1} |B_n(t)| \sup_{h^{2/3} \le u \le 1/2} \sup_{0 \le y \le h} h \frac{(\xi(y))^{\alpha}}{f(Q(\xi(y)))} \frac{u^{1-\alpha}}{Q(u)} \left(\frac{u}{\xi(y)}\right)^{\alpha} u^{-1}$$

$$\overset{a.s.}{=} h^{1/3} O((\log\log n)^{1/2})$$

as a consequence of (2.15), (13.27) and (13.29). Similarly,

$$\Delta_1^{(6)}(n,h) \le \sup_{0 \le t \le 1} |B_n(t)| \sup_{h^{2/3} \le u \le 1/2} h u^{-1} \frac{u^{1-\alpha}}{Q(u)}$$

$$\times \sup_{0 \le y \le h} \frac{(\xi(y))^{\alpha}}{f(Q(\xi(y))} \frac{Q(u+y) - Q(0)}{Q(u+y)}$$

$$\stackrel{a.s.}{=} h^{1/3}\, O((\log\log n)^{\frac{1}{2}}).$$

So, collecting $\Delta_1^{(1)}, \dots, \Delta_1^{(6)}$, we get

$$\Delta_1(n,h) \stackrel{a.s.}{=} h^{1/3} O((\log n)^{1/3}). \tag{13.53}$$

Now the form of $\tilde{\Gamma}_n$ in (13.51) and easy algebraic manipulations yield

$$\begin{aligned}
\Delta_2(n,h) \le & \sup_{1/2\le u\le 1-h}\ \sup_{0\le y\le h} |B_n(u+y) - B_n(u)| \\
& + \left|\int_0^1 \frac{B_n(t)}{f(Q(t))}\,dt\right| \left\{ 2 \sup_{1-h^{2/3}\le u\le 1} \frac{1-L_F(u)}{Q(u)} \right. \\
& \left. + \sup_{1/2\le u\le 1-h^{2/3}}\ \sup_{0\le y\le h} \left| \frac{1-L_F(u)}{Q(u)} - \frac{1-L_F(u+y)}{Q(u+y)} \right| \right\} \\
& + 2 \sup_{1-h^{2/3}\le u\le 1} \frac{1}{Q(u)} \left| \int_u^1 \frac{B_n(t)}{f(Q(t))}\,dt \right| \\
& + \sup_{1/2\le u\le 1-h^{2/3}}\ \sup_{0\le y\le h} \left(\frac{1}{Q(u)} - \frac{1}{Q(u+y)}\right) \left| \int_u^1 \frac{B_n(t)}{f(Q(t))}\,dt \right| \\
& + \sup_{1/2\le u\le 1-h^{2/3}}\ \sup_{0\le y\le h} \frac{1}{Q(u+y)} \left| \int_u^{u+y} \frac{B_n(t)}{f(Q(t))}\,dt \right| \\
= & \ \Delta_2^{(1)}(n,h) + \dots + \Delta_2^{(5)}(n,h),
\end{aligned}$$

where

$$\Delta_2^{(1)}(n,h) \stackrel{a.s.}{=} h^{\frac{1}{2}} O((\log n)^{\frac{1}{2}})$$

by Lemma 5.4. By (13.30) and (13.33)

$$\sup_{1-h^{2/3}\le u\le 1} \frac{1-L_F(u)}{Q(u)} \le C_{31}\, h^{2/3},$$

and

$$\begin{aligned}
& \sup_{0\le y\le h} \left| \frac{1-L_F(u)}{Q(u)} - \frac{1-L_F(u+y)}{Q(u+y)} \right| \\
& \le \frac{1-L_F(u)}{Q(u)} \sup_{0\le y\le h} \frac{Q(u+y)-Q(u)}{Q(u+y)} + \sup_{0\le y\le h} \frac{1}{Q(u+y)} \int_u^{u+y} Q(t)\,dt \\
& \le \frac{1-L_F(u)}{Q(u)Q(u)}\, h \sup_{0\le y\le h} \frac{(1-\xi(y))^{\beta}}{f(Q(\xi(y)))} \left(\frac{1-u}{1-\xi(y)}\right)^{\beta} (1-u)^{-\beta} + h
\end{aligned}$$

$$\le h\{1 + C_{32} \frac{(1-L_F(u))(1-u)^{-\beta}}{Q^2(u)}\}$$

$$\le C_{33} h$$

by (13.30) and (13.33), as a separate inspection of the three cases $\beta < 1$, $\beta = 1$ and $\beta > 1$ shows. Therefore,

$$\Delta_2^{(2)}(n,h) \overset{a.s.}{=} h^{2/3} O((\log\log n)^{\frac{1}{2}}).$$

Next, by condition (13.28) and (2.9)

$$\left|\int_u^1 \frac{B_n(t)}{f(Q(t))} dt\right| \le \sup_{0\le t\le 1} \frac{|B_n(t)|}{(1-t)^{\frac{1}{2}-\delta}} C_{34} \int_u^1 (1-t)^{\frac{1}{2}-\beta-\delta} dt$$

$$\overset{a.s.}{=} (1-u)^{3/2-\beta-\delta} O((\log\log n)^{\frac{1}{2}}), \quad 1/2 \le u \le 1,$$

for any $\delta > 0$. Hence, again by (13.30),

$$\Delta_2^{(3)}(n,h) \overset{a.s.}{=} h^{1/3-\delta} O((\log\log n)^{\frac{1}{2}})$$

for any $\delta > 0$. Also, since

$$\sup_{0\le y\le h} \left(\frac{1}{Q(u)} - \frac{1}{Q(u+y)}\right) \le h \frac{1}{Q^2(u)} \sup_{0\le y\le h} \frac{(1-\xi(y))^{\beta}}{f(Q(\xi(y)))}\left(\frac{1-u}{1-\xi(y)}\right)^{\beta}(1-u)^{-\beta}$$

$$\le C_{32} h (Q(u))^{-2}(1-u)^{-\beta},$$

(13.30) gives

$$\Delta_2^{(4)}(n,h) \overset{a.s.}{=} h^{1/3-\delta} O((\log\log n)^{\frac{1}{2}})$$

for any $\delta > 0$. Finally, exactly as in the last inequality,

$$\Delta_2^{(5)}(n,h) \le C_{32} \sup_{0\le t\le 1} |B_n(t)| \sup_{1/2\le u\le 1-h^{2/3}} h \frac{(1-u)^{-\beta}}{Q(u)}$$

$$\overset{a.s.}{=} h^{1/2} O(\log\log n)^{\frac{1}{2}}).$$

Thus, altogether,

$$\Delta_2(n,h) \overset{a.s.}{=} h^{1/3-\delta} O((\log\log n)^{\frac{1}{2}}).$$

This and (13.53) now give Lemma 13.11.

At last we are in the position to state and prove a strong approximation theorem for the Goldie concentration process.

THEOREM 13.12. _If the density function_ $f = F'$ _is positive on the open support of_ F _and conditions_ (13.27) and (13.28) hold, then

$$\Delta_n^{(11)} = \sup_{0\le y\le 1} |c_n(y) - \Psi_n(y)| \overset{a.s.}{=} O(n^{-\tau}) ,$$

where $\tau < \min(\tau_1, \tau_2, 1/6)$ with

$$\tau_1 = \begin{cases} \min(\frac{1}{2}, \frac{3}{2} - \beta) , & \text{if } Q(0) > 0, \\ \dfrac{\min(\frac{1}{2}, \frac{3}{2} - \beta)}{3 - 2\alpha} , & \text{if } Q(0) = 0 \end{cases} \quad \text{and} \quad \tau_2 = \begin{cases} \dfrac{1}{2(2\beta+1)} , & \text{if } \beta < 1, \\ \dfrac{3-2\beta}{4(\beta^2-1)+6} , & \text{if } \beta \ge 1. \end{cases}$$

Proof. We start out from the three-term bound of $\Delta_n^{(11)}$ at the beginning of the proof of Theorem 13.5. By (2.1),

$$\sup_{0\le u\le 1} |B_n(C_n^{-1}(u)) - \alpha_n(C_n^{-1}(u))| \overset{a.s.}{=} O(n^{-\frac{1}{2}}(\log n)^2).$$

Almost as in the proof of Lemma 13.1

$$\begin{aligned} \sup_{0\le u\le 1} |C_n^{-1}(u) - L_F^{-1}(u)| &= \sup_{0\le u\le 1} |C_n^{-1}(L_F(u)) - u| \\ &\le \sup_{0\le u\le 1} |L_F^{-1}(C_n(u) - u| \qquad (13.54) \\ &= \sup_{0\le u\le 1} n^{-\frac{1}{2}} |\gamma_u(u)| \\ &\overset{a.s.}{=} O((n^{-1} \log\log n)^{\frac{1}{2}}) \end{aligned}$$

as a consequence of the strong approximation in Lemma 13.10 and the fact that

$$\sup_{0\le u\le 1} |\frac{\mu}{Q(u)} \Gamma_n^*(u)| \overset{a.s.}{=} O((\log\log n)^{\frac{1}{2}}). \qquad (13.55)$$

The latter follows from inequalities implied by our conditions and from (2.9) in the usual way (cutting at 1/2, integration by parts, and so on). Hence by (13.54) and Lemma 5.4,

$$\sup_{0\le u\le 1} |B_n(C_n^{-1}(u)) - B_n(L_F^{-1}(u))| \overset{a.s.}{=} O((n^{-1}\log\log n)^{1/4}(\log n)^{1/2}).$$

Thus our only remaining task is to show that

$$A_{1n} = \sup_{0\le u\le 1} |n^{\frac{1}{2}}\{C_n^{-1}(L_F(u)) - u\} + \tilde{\Gamma}_n(u)|$$

is of the order $n^{-\tau}$. Using (13.12) and (13.24),

$$A_{1n} \le \sup_{0\le u\le 1} |\tilde{\Gamma}_n(u) - \gamma_n(C_n^{-1}(L_F(u)))|$$

$$+2 \sup_{0\le u\le 1} |\tilde{\Gamma}_n(u) - \gamma_n(u)|$$

$$+ \sup_{0\le u\le 1-\frac{1}{n}} \sup_{0\le y\le\frac{1}{n}} |\tilde{\Gamma}_n(u+y) - \tilde{\Gamma}_n(u)|$$

$$\le 3 \sup_{0\le u\le 1} |\tilde{\Gamma}_n(u)-\gamma_n(u)|$$

$$+ \sup_{0\le u\le 1} |\tilde{\Gamma}_n(u) - \tilde{\Gamma}_n(C_n^{-1}(L_F(u)))|$$

$$+ \sup_{0\le u\le 1-\frac{1}{n}} \sup_{0\le y\le\frac{1}{n}} |\tilde{\Gamma}_n(u+y) - \tilde{\Gamma}_n(u)|$$

$$= 3A_{2n} + A_{17n} + A_{18n} ,$$

where

$$A_{2n} \overset{a.s.}{=} O(n^{-\rho})$$

by Lemma 13.10. For A_{17n}, (13.54) and Lemma 13.11 give that

$$A_{17n} \overset{a.s.}{=} O(n^{-\frac{1}{6}+\delta})$$

for any $\delta > 0$. And lastly, Lemma 13.11 directly gives that

$$A_{18n} \overset{a.s.}{=} O(n^{-\frac{1}{3}+\delta})$$

for any $\delta > 0$. Now the definition of ρ in Lemma 13.10 gives the theorem.

The form of the approximating processes Ψ_n in (13.9) shows that we have a loglog law for Ψ_n if and only if we have one for Φ_n, for L_F^{-1} is a homeomorphism on $[0,1]$. For the Kiefer term of Φ_n in (13.10) we know (2.15) exactly, while for $\tilde{\Gamma}_n = \mu\Gamma_n^*/Q$ it is easy to show that, as a more precise form of (13.55), we have

$$\limsup_{n\to\infty} (\log\log n)^{-\frac{1}{2}} \sup_{0\le u\le 1} |\tilde{\Gamma}_n(u)| \le K(F) - \frac{1}{\sqrt{2}} \quad \text{a.s.},$$

where, with $h(y) = (y(1-y)\log\log \frac{1}{y(1-y)})^{\frac{1}{2}}$, $y \in (0,1)$, $K(F)$ is given by

$$K(F) = 2 \sup_{0\le u\le 1} \frac{L_F(u)}{Q(u)} \int_0^1 h(y)\,dQ(y) + 2\int_0^1 \frac{h(y)}{Q(y)}\,dQ(y) + \frac{2}{\sqrt{2}} .$$

Therefore Theorem 13.12 implies the following uniform consistency result.

COROLLARY 13.13. <u>Under the conditions of Theorem</u> 13.12

$$\limsup_{n\to\infty} \left(\frac{n}{\log\log n}\right)^{\frac{1}{2}} \sup_{0\le u\le 1} |L_n^{-1}(u)-L_F^{-1}(u)| \le K(F)$$

<u>almost surely</u>.

14. DISCUSSION OF RESULTS ON THE GOLDIE CONCENTRATION PROCESS

1) If the lower endpoint of the support of F is positive, that is, $t_F > 0$, then our condition (13.11) for the weak approximation in Theorem 13.5 is trivially satisfied. Indeed, $\lim_{u\to 0} Q(u) = t_F = \lim_{u\to 0} Q(u/\lambda)$ for each $\lambda > 1$, and therefore condition (13.11) takes the form

$$\limsup_{u\to 0} \frac{Q(u)q(u)}{Q(u/\lambda)} = \lim_{u\to 0} q(u) = 0$$

in this case.

2) Suppose that $EX^2 < \infty$ and Q is continuous on $(0,1)$. If $t_F > 0$ then Goldie (1977) also proved the weak convergence of $c_n(\cdot)$ to $\Psi(\cdot)$ relative to the supremum norm. When $t_F = 0$ he also used an extra variation condition on Q around zero. In his Proposition 8 he gives eight equivalent forms (C8-C15) of his variation condition, the last one of which is obviously equivalent to the following condition: There exist positive constants C, a and u_o such that

(14.1) $$\frac{Q(s)}{Q(t)} \le C\left(\frac{s}{t}\right)^a \quad \text{whenever} \quad 0 < t \le s \le u_o.$$

This condition is not satisfied automatically when $t_F > 0$ and this is why Goldie has to separate the easy case $t_F > 0$ from the hard case $t_F = 0$. Note also that Goldie's lower variation condition (14.1) is exactly parallel to his upper variation condition (11.2) for the Lorenz process. Of course Goldie's condition (14.1) implies that for any $\lambda \ge 1$

$$\frac{Q(u)}{Q(u/\lambda)} \le C\lambda^a \quad \text{whenever} \quad 0 < u \le u_o ,$$

and *a fortiori* that

$$\limsup_{u\to 0} \frac{Q(u)}{Q(u/\lambda)} < \infty$$

for each $\lambda > 1$. This latter condition is stronger than our variation condition (13.11) where the extra O'Reilly function q can only help since $\lim_{u\to 0} q(u) = 0$. Thus our variation condition (13.11) for the Goldie process is much weaker than that of Goldie's one. For further comparison of the results for the Lorenz and the concentration processes we note that in the proof of Theorem 10.2 we inferred more from the (necessary) condition of a finite variance than Goldie did, and this is why we were able to get free from any upper variation condition. On the other hand, the condition $EX^2 < \infty$ does not provide any information about the behaviour of Q near zero. This is why we needed a lower variation condition there, in Theorem 13.5, although a weaker one than what Goldie required. We don't know if the weak convergence result for the concentration process is true without any lower variation condition on Q.

Again, just as for the Lorenz process, Goldie also proved the weak convergence of c_n in Skorohod's M_1 topology if $EX^2 < \infty$ and if his condition (14.1) holds true, but Q in his case is possibly discontinuous. He has a general representation of the limit process $\Psi(\cdot)$ which is different from ours in (13.9) and (13.10). If Q is continuous, then his representation can be brought into the form of our Ψ.

3) In his probing review of Goldie (1977), which has directed our attention to Goldie's paper, Wellner (1979) remarks that in the case when Q is continuous Goldie's weak convergence theorem for his concentration process $c_n(\cdot)$ "may be obtained quite simply" from the weak convergence result for the Lorenz process $\ell_n(\cdot)$ "by use of results of Vervaat (1972) on the convergence of inverse processes". Wellner (1979) also notes that the limit process of $c_n(\cdot)$ must be the $\Psi(\cdot)$ process in (13.9) and (13.10), that is, $\Psi(\cdot)$ as in (14.2) below. As we have seen, Wellner's remark is completely right concerning the form of the limiting process and the general message of his remark has helped us in finding our proof for Theorem 13.5. On the other hand, we do not see how Vervaat's (1972) results could be directly used for the derivation of the weak convergence of c_n from that of ℓ_n. In order to be in Vervaat's situation, one must transform the process c_n in such a way that the theoretical function $L_F^{-1}(u)$ in it be simply u. This is indeed what is basically achieved in (13.12) by the introduction of the process $\gamma_n(u)$. Then Vervaat's Theorem 1 implies that the weak convergence of

$$\gamma_n(u) = n^{\frac{1}{2}}\{C_n^{-1}(L_F(u))-u\}$$

to $-\mu\Gamma^*(u)/Q(u)$ is equivalent to the weak convergence of

$$n^{\frac{1}{2}}\{L_F^{-1}(C_n(u))-u\}$$

to $\mu\Gamma^*(u)/Q(u)$. Of course the weak convergence of $\gamma_n(\cdot)$ does not follow from that of $\ell_n(\cdot)$, or, what is nearly the same, from that of $\ell_n^*(\cdot) = n^{\frac{1}{2}}\{C_n(\cdot)-L_F(\cdot)\}$, and the main body of our proof is indeed the proof of such a statement for $\gamma_n(\cdot)$ in (13.13). By specifying Vervaat's theorem to the present situation we could only save (13.22) and the short and suggestive proof of (13.23). The reason why we did not do so is that we wanted to show clearly the steps which must have been strengthened later on in order to obtain the strong approximation in Theorem 13.12. Thus the strong form of (13.22) appears in the estimation of A_{1n} in the proof of Theorem 13.12, and the proof of (13.23) clearly suggested that we should determine an order for the modulus of continuity of $\tilde{\Gamma}_n(u) = \mu\Gamma_n^*(u)/Q(u)$. We note also that the weak convergence of $\ell_n(\cdot)$ required harder work with an extra condition in the vicinity of 1, while that of c_n required the same around zero.

4) The limit process of the Goldie concentration processes $c_n(\cdot)$ is

$$(14.2) \quad \Psi(u) = \Psi_F(u) = \frac{1}{Q(L_F^{-1}(u))}\left\{u\int_0^1 B(y)dQ(y) - \int_0^{L_F^{-1}(u)} B(y)dQ(y)\right\},$$

$0 \le u \le 1$. Its covariance is

$$\sigma_5(s,t) = E\Psi_F(s)\Psi_F(t) = \frac{\mu^2\sigma_4(L_F^{-1}(s), L_F^{-1}(t))}{Q(L_F^{-1}(s))Q(L_F^{-1}(t))}, \quad 0 \le s, t \le 1,$$

where $\sigma_4(\cdot,\cdot)$ is the limiting covariance function of the Lorenz process as given in Section 12. Hence the variance function is

$$\sigma_5^2(t) = \sigma_5(t,t) = E\Psi_F^2(t) = \frac{\mu^2\sigma_4^2(L_F^{-1}(t))}{Q^2(L_F^{-1}(t))}.$$

Again, we could not compute the distribution of $\phi(\Psi_F(\cdot))$ for any reasonable functional ϕ, and for any underlying distribution function F. See, however, Theorem 17.1.

The variance function σ_5^2 can now be written as

$$\sigma_5^2(t) = \frac{1}{Q^2(L_F^{-1}(t))}\left\{2\int_0^{L_F^{-1}(t)} (1-u)\{Q(u)-N_F^1(u)\}dQ(u)[1-2t]\right.$$

$$+ 2t^2\int_0^1 (1-u)\{Q(u) - N_F^1(u)\}dQ(u)$$

$$\left. - 2t[N_F^1(1)-N_F^1(L_F^{-1}(t))][Q(L_F^{-1}(t))-N_F(L_F^{-1}(t))]\right\}$$

where

$$N_F^1(s) = \int_0^s (1-y)dQ(y).$$

Replacing Q by Q_n of (8.5) and L_F^{-1} by L_n^{-1} we obtain $\sigma_{5n}^2(t)$, an estimator of $\sigma_5^2(t)$. For any fixed $u \in (0,1)$ we have the following.

COROLLARY 14.1. <u>Under the conditions of Theorem</u> 13.5

$$\lim_{n\to\infty} \mathrm{pr}\left\{\frac{c_n(u)}{\sigma_5(u)} \le x\right\} = \Phi(x), \quad -\infty < x < \infty,$$

<u>and</u>

$$\lim_{n\to\infty} \mathrm{pr}\{L_n^{-1}(u) - x\,\frac{\sigma_{5n}(u)}{\sqrt{n}} \le L_F^{-1}(u) \le L_n^{-1}(u) + \frac{\sigma_{5n}(u)}{\sqrt{n}}\} = 2\Phi(x) - 1,$$

$-\infty < x < \infty$, <u>where</u> Φ <u>is the standard normal distribution function</u>.

5) Goldie (1977) has not considered the inverse process to the unscaled Lorenz process. This process is perhaps indeed not important in itself, but we shall need it in the next section. Define

$$G_n^{-1}(x) = \begin{cases} 0\,, & \text{if } 0 \le x < n^{-1}X_{1:n}\,, \\ \frac{k-1}{n}\,, & \text{if } n^{-1}\sum_{i=1}^{k-1} X_{i:n} \le x < n^{-1}\sum_{i=1}^{k} X_{i:n},\ 2 \le k \le n, \\ 1\,, & \text{if } x = \overline{X}_n\,, \end{cases}$$

and

$$G_F^{-1}(x) = \inf\{u : G_F(u) > x\},\quad 0 \le x \le G_F(1) = \mu\,,$$

where

$$G_F(u) = \int_0^u Q(y)\,dy\,,\quad 0 \le u \le 1.$$

Then the inverse unscaled Lorenz process is

$$g_n^{-1}(x) = n^{\frac{1}{2}}\{G_n^{-1}(x) - G_F^{-1}(x)\},\quad 0 \le x \le \mu_n$$

where $\mu_n = \min(\mu, \overline{X}_n)$, and

$$g_n^{-1}(x) = g_n^{-1}(\mu_n)\,,\quad \mu_n \le x \le \mu\,.$$

Introduce also the Gaussian processes

$$\Gamma_n^{-1}(x) = -\frac{dG_F^{-1}(y)}{dy}\,\Gamma_n(y)\Big|_{y = G_F^{-1}(x)}$$

$$= -\frac{1}{Q(G_F^{-1}(x))}\int_0^{G_F^{-1}(x)} B_n(y)\,dQ(y),\quad 0 \le x \le \mu\,,$$

where $E\Gamma_n^{-1}(x) = 0$ and the covariance function is

$$\sigma_6(s,t) = E\Gamma_n^{-1}(s)\Gamma_n^{-1}(t) = E\Gamma_F^{-1}(s)\Gamma_F^{-1}(t)$$

$$= \frac{\sigma_3(G_F^{-1}(s), G_F^{-1}(t))}{Q(G_F^{-1}(s))\,Q(G_F^{-1}(t))}\,,$$

where $\sigma_3(\cdot,\cdot)$ is the covariance function of $\Gamma_F(\cdot)$ given in Section 12. Note that

$$\Gamma_n^{-1}(\mu) = \frac{1}{Q(1)} \int_0^1 B_n(y)\,dQ(y) = 0 \quad \text{a.s.}$$

if and only if $Q(1) = \infty$, that is, if and only if $T_F = \infty$. The proofs of the following three results are obvious simplifications of the corresponding proofs of Theorems 13.2, 13.5 and 13.12 after noticing that

$$P\{\sup_{0 \le x \le \bar{X}_n} |G_n^{-1}(x) - E_n(\inf\{u : \int_0^u Q(y)\,dE_n(y) > x\})| = 0\} = 1,$$

a parallel observation to (13.2).

THEOREM 14.1. (i) <u>If</u> $\mu < \infty$ <u>then</u>

$$\sup_{0 \le x \le \mu_n} |G_n^{-1}(x) - G_F^{-1}(x)| \xrightarrow{a.s.} 0.$$

(ii) <u>If the conditions of Theorem</u> 13.5 <u>are satisfied then</u>

$$\sup_{0 \le x \le \mu} |g_n^{-1}(x) - \Gamma_n^{-1}(x)| \xrightarrow{P} 0 .$$

(iii) <u>If the conditions of Theorem</u> 13.12 <u>are satisfied then</u>

$$\sup_{0 \le x \le \mu} |g_n^{-1}(x) - \Gamma_n^{-1}(x)| \overset{a.s.}{=} O(n^{-\tau}) ,$$

<u>where</u> τ <u>is as in Theorem</u> 13.12.

15. FURTHER DIVERSITY AND CONCENTRATION PROCESSES

Here we use the term "diversity" as the opposite to "concentration", that is, as the economic-theoretical synonym of "inequality". In this terminology the basic empirical diversity process is the empirical Lorenz process $\ell_n(\cdot)$ and the basic empirical concentration process is the empirical Goldie concentration process $c_n(\cdot)$. In their recent survey Patil and Taillie (1982) mention a few more indices as possible competitors of the Gini coefficient of inequality, which are also used in economics although not as frequently as Gini's mean difference. Sugihara (1982) as one of the discussants of the paper by Patil and Taillie (1982), although from the point of view of a researcher of ecologic diversity, but possibly validly also from the point of view of economics, points out that "it seems premature at this early stage to restrict attention to indices and risk losing information by condensing the complete abundance vector into a single statistic. It would be more prudent in these initial stages to consider the complete abundance vector as represented by ranked abundance plots...". These comments have led us to introduce below some more empirical diversity processes, corresponding to indices mentioned by Patil and Taillie (1982), and their concentration inverses. Although some of these processes are formally more general or seem more complicated than the original Lorenz process, it will become clear that the Lorenz process is indeed the basic process, at least within the class of processes introduced below, because the formally more general porcesses as mathematical objects may be treated as special cases of the Lorenz process and its inverse.

15.1. Empirical Lorenz processes of order ν.

Let F be a continuous life distribution function $(t_F \geq 0)$ as before, and let ν be any nonzero fixed real number. If $X_1,\ldots,X_n$ is a sample from F, consider the random variables

$$Y_k = Y_k(\nu) = X_k^{\nu}, \quad k = 1,\ldots,n,$$

and let $Y_{1:n}(\nu) \leq \ldots \leq Y_{n:n}(\nu)$ be the ordered sample corresponding to $Y_1,\ldots,Y_n$. Motivated by a Rényi (1961) type index considered by Patil and Taillie (1982) and formerly by Hill (1972), consider the unscaled empirical Lorenz curve of order ν defined as

$$G_n^{(\nu)}(u) = \begin{cases} \frac{1}{n} \sum_{i=1}^{[nu]+1} Y_{i:n}(\nu), & \text{if } 0 \leq u < 1, \\ \frac{1}{n} \sum_{i=1}^{n} Y_{i:n}(\nu), & \text{if } u = 1. \end{cases}$$

Of course,

$$G_n^{(\nu)}(1) = \frac{1}{n}\sum_{i=1}^{n} X_i^{\nu} ,$$

the empirical moment of order ν , and if $\nu > 0$ then

$$G_n^{(\nu)}(u) = \frac{1}{n}\sum_{i=1}^{[nu]+1} X_{i:n}^{\nu} , \quad 0 \le u < 1,$$

and if $\nu < 0$ then

$$G_n^{(\nu)}(u) = \frac{1}{n}\sum_{i=1}^{[nu]+1} X_{n-i+1:n}^{\nu} , \quad 0 \le u < 1.$$

The corresponding theoretical quantity is

$$G_F^{(\nu)}(u) = \int_0^u Q_\nu(y)\,dy, \quad 0 \le u \le 1,$$

where $Q_\nu(y)$ is the common quantile function of $X_1^\nu, X_2^\nu, \ldots,$ that is,

$$Q_\nu(y) = \begin{cases} Q^\nu(y) , & \text{if } \nu > 0 \\ Q^\nu(1-y) , & \text{if } \nu < 0. \end{cases}$$

Then the unscaled empirical Lorenz process of order ν ,

$$g_n^{(\nu)}(u) = n^{\frac{1}{2}}\{G_n^{(\nu)}(n) - G_F^{(\nu)}(u)\}, \ 0 \le u \le 1,$$

may be looked upon as the ordinary unscaled empirical Lorenz process of the variables $X_1^\nu, X_2^\nu, \ldots .$ The copies of the limiting zero-mean Gaussian porcesses are

$$\Gamma_n^{(\nu)}(u) = \int_0^u B_n(y)\,dQ_\nu(y), \quad 0 \le u \le 1,$$

for which

$$\Gamma_n^{(\nu)}(u) = \begin{cases} \nu \int_0^u B_n(y) Q^{\nu-1}(y)\,dQ(y) , & \text{if } \nu > 0, \\ \nu \int_0^u B_n(y) Q^{\nu-1}(1-y)\,dQ(1-y) , & \text{if } \nu < 0. \end{cases}$$

We also introduce all the inverse quantities as

$${}_\nu G_n^{-1}(x) = \begin{cases} 0 \ , & \text{if } 0 \le x < n^{-1} Y_{1:n} \\ \frac{k-1}{n} \ , & \text{if } n^{-1}\sum_{i=1}^{k-1} Y_{i:n} \le x < n^{-1}\sum_{i=1}^{k} Y_{i:n}, \ 2 \le k \le n, \\ 1 \ , & \text{if } x = n^{-1}\sum_{i=1}^{n} Y_i \ , \end{cases}$$

$${}_{\nu}G_F^{-1}(x) = \inf\{u : G_F^{(\nu)}(u) > x\}, \quad 0 \le x \le \mu^{(\nu)} = G_F^{(\nu)}(1) = EX^{\nu},$$

$${}_{\nu}g_n^{-1}(x) = \begin{cases} n^{\frac{1}{2}}\{{}_{\nu}G_n^{-1}(x) - {}_{\nu}G_F^{-1}(x)\}, & 0 \le x \le \mu_n^{(\nu)}, \\ {}_{\nu}g_n^{-1}(\mu_n^{(\nu)}), & \mu_n^{(\nu)} \le x \le \mu^{(\nu)}, \end{cases}$$

where $\mu_n^{(\nu)} = \min(n^{-1}\sum_{i=1}^{n} X_i^{\nu}, \mu^{(\nu)})$, and

$${}_{\nu}\Gamma_n^{-1}(x) = -\frac{d\,{}_{\nu}G_F^{-1}(y)}{dy}\,\Gamma_n^{(\nu)}(y)\Big|_{y = {}_{\nu}G_F^{-1}(x)}$$

$$= -\frac{1}{Q_{\nu}({}_{\nu}G_F^{-1}(x))}\int_0^{{}_{\nu}G_F^{-1}(x)} B_n(y)\,dQ_{\nu}(y), \quad 0 \le x \le \mu^{(\nu)}.$$

Both $\Gamma_n^{(\nu)}(u)$ and ${}_{\nu}\Gamma_n^{-1}(x)$ have zero mean, and their covariance functions are easily obtained. The following results are simple consequences of Theorems 10.1, 10.2, 10.3 and 14.1. When $\nu = 1$ these results reduce to the corresponding statements of these theorems.

THEOREM 15.1. (i) *If* $\mu^{(\nu)} = EX^{\nu} < \infty$, *then*

$$\sup_{0 \le u \le 1} |G_n^{(\nu)}(u) - G_F^{(\nu)}(u)| \xrightarrow{a.s.} 0$$

and

$$\sup_{0 \le x \le \mu_n^{(\nu)}} |{}_{\nu}G_n^{-1}(x) - {}_{\nu}G_F^{-1}(x)| \xrightarrow{a.s.} 0 .$$

(ii) *Suppose that* Q *is continuous on* $(0,1)$ *and* $EX^{2\nu} < \infty$. *Then*

$$\sup_{0 \le u \le 1} |g_n^{(\nu)}(u) - \Gamma_n^{(\nu)}(u)| \xrightarrow{P} 0.$$

If for each $\lambda > 1$

$$\limsup_{u \to 0} \left(\frac{Q(u)}{Q(u/\lambda)}\right)^{\nu} q(u) < \infty \quad \textit{when} \quad \nu > 0,$$

or

$$\limsup_{u \to 0} \left(\frac{Q(1-u)}{Q(1-\frac{u}{\lambda})}\right)^{\nu} q(u) < \infty \quad \textit{when} \quad \nu < 0,$$

hold true for some O'Reilly weight function q , *then*

$$\sup_{0\le x\le\mu^{(\nu)}} |{}_\nu g_n^{-1}(x) - {}_\nu\Gamma_n^{-1}(x)| \xrightarrow{P} 0 .$$

(iii) _Suppose that the density function_ $f = F'$ _is positive on the open support of_ F . _If_

$$\sup_{0<u<1} \frac{u^\alpha(1-u)^\beta Q^{\nu-1}(u)}{f(Q(u))} < \infty \quad \text{when} \quad \nu > 0,$$

or

$$\sup_{0<u<1} \frac{u^\alpha(1-u)^\beta Q^{\nu-1}(1-u)}{f(Q(1-u))} < \infty \quad \text{when} \quad \nu < 0$$

with $0 \le \alpha < 3/2$, $0 \le \beta < 3/2$, _then_

$$\sup_{0\le u\le 1} |g_n^{(\nu)}(u) - \Gamma_n^{(\nu)}(u)| \overset{a.s.}{=} O(n^{-\tau})$$

where τ _is as in Theorem_ 10.3. _If_

$$0 < \lim_{u\to 0} \frac{u^\alpha Q^{\nu-1}(u)}{f(Q(u))} < \infty \quad \text{and} \quad 0 < \lim_{u\to 1} \frac{(1-u)^\beta Q^{\nu-1}(u)}{f(Q(u))} < \infty$$

when $\nu > 0$, _or_

$$0 < \lim_{u\to 0} \frac{u^\alpha Q^{\nu-1}(1-u)}{f(Q(1-u))} < \infty \quad \text{and} \quad 0 < \lim_{u\to 1} \frac{(1-u)^\beta Q^{\nu-1}(1-u)}{f(Q(1-u))} < \infty$$

when $\nu < 0$, _where_ $0 \le \alpha,\beta < 3/2$, _then_

$$\sup_{0\le x\le\mu^{(\nu)}} |{}_\nu g_n^{-1}(x) - {}_\nu\Gamma_n^{-1}(x)| \overset{a.s.}{=} O(n^{-\tau}),$$

where τ _is as in Theorem_ 13.12.

15.2. _Empirical Shannon processes_

Given a sample $X_1,\dots,X_n$ from a continuous life distribution F, consider the random variables

$$Z_k = X_k \log X_k, \quad k=1,\dots,n,$$

and let $Z_{1:n} \le \dots \le Z_{n:n}$ be the order statistics of the transformed sample $Z_1,\dots,Z_n$. We define the sample Shannon function as

$$\bar{S}_n(u) = \begin{cases} \frac{1}{n}\sum_{k=1}^{[nu]+1} Z_{k:n}, & 0 \le u \le 1, \\ \bar{Z}_n = \frac{1}{n}\sum_{k=1}^{n} X_k \log X_k, & u = 1, \end{cases}$$

with theoretical counterpart

(15.1) $$\bar{S}_F(u) = \int_0^u \bar{Q}(y)\,dy, \quad 0 \le u \le 1,$$

where

$$\bar{Q}(y) = \inf\{u : \bar{F}(u) > y\}$$

is the quantile function of $X \log X$, i.e.,

$$\bar{F}(u) = \mathrm{pr}\{X \log X \le u\}.$$

The empirical Shannon process is then

$$\bar{s}_n(u) = n^{\frac{1}{2}}\{\bar{S}_n(u) - \bar{S}_F(u)\}, \quad 0 \le u \le 1.$$

Since $\bar{S}_n$ is not usually non-decreasing, its inverse should be defined with caution. A natural definition is

$$\bar{S}_n^{-1}(x) = \begin{cases} 0, & -e^{-1} \le x < n^{-1}Z_{1:n}, \\ \frac{k-1}{n}, & n^{-1}\sum_{i=1}^{k-1} Z_{i:n} \le x < n^{-1}\sum_{i=1}^{k} Z_{i:n}, \quad 2 \le k \le n, \\ 1, & x = \bar{Z}_n. \end{cases}$$

Note that this inverse does not satisfy (1.2). Again, $\bar{S}_F(x)$ is not non-decreasing in general, and we define its inverse as

$$\bar{S}_F^{-1}(x) = \inf\{u \ge 0 : \bar{S}_F(u) + \frac{u}{e} > x + \frac{1}{e}\}, \quad -\frac{1}{e} \le x \le \bar{\mu}$$

where

$$\bar{\mu} = EX \log X.$$

Now the inverse empirical Shannon process, or the empirical Shannon concentration process is

$$\bar{s}_n^{-1}(x) = \begin{cases} n^{\frac{1}{2}}\{\bar{S}_n^{-1}(x) - \bar{S}_F^{-1}(x)\}, & -\frac{1}{e} \le x \le \bar{\mu}_n, \\ \bar{s}_n^{-1}(\bar{\mu}_n), & \bar{\mu}_n \le x \le \bar{\mu}. \end{cases}$$

where

$$\bar{\mu}_n = \min(\bar{Z}_n, \bar{\mu}) = \min(\frac{1}{n}\sum_{i=1}^{n} X_i \log X_i, \bar{\mu}).$$

The corresponding approximating sequences of the limit processes are

(15.2) $$\bar{\Gamma}_n(u) = \int_0^u B_n(y)\,d\bar{Q}(y), \quad 0 \le u \le 1,$$

and

$$\bar{\Gamma}_n^{-1}(x) = -\frac{d\bar{S}_F^{-1}(y)}{dy}\,\bar{\Gamma}_n(y)\Big|_{y=\bar{S}_F^{-1}(x)}$$

$$= -\frac{1}{\bar{Q}(\bar{S}_F^{-1}(x)) + \frac{1}{e}}\int_0^{\bar{S}_F^{-1}(x)} B_n(y)\,d\bar{Q}(y), \quad -\frac{1}{e} \le x \le \bar{\mu}.$$

If we apply Theorems 10.1, 10.2, 10.3 and 14.1 to the variables $Z_k' = Z_k + e^{-1}$, $k=1,2,\ldots$, with all the corresponding quantities, we arrive at the following results.

THEOREM 15.2. (i) *If* $\bar{\mu} = EX \log X < \infty$, *then*

$$\sup_{0\le u\le 1} |\bar{S}_n(u) - \bar{S}_F(u)| \xrightarrow{a.s.} 0$$

and

$$\sup_{0\le x\le \bar{\mu}_n} |\bar{S}_n^{-1}(x) - \bar{S}_F^{-1}(x)| \xrightarrow{a.s.} 0.$$

(ii) *Suppose that* $\bar{Q}$ *is continuous on* $(0,1)$ *and* $E(X \log X)^2 < \infty$. *Then*

$$\sup_{0\le u\le 1} |\bar{s}_n(u) - \bar{\Gamma}_n(u)| \xrightarrow{P} 0.$$

If for each $\lambda > 1$

$$\limsup_{u\to 0} \frac{\bar{Q}(u) + e^{-1}}{\bar{Q}(u/\lambda)+e^{-1}}\, q(u) < \infty$$

for some O'Reilly weight function q, *then*

$$\sup_{-e^{-1}\le x\le \bar{\mu}} |\bar{s}_n^{-1}(x) - \bar{\Gamma}_n^{-1}(x)| \xrightarrow{P} 0.$$

(iii) *Suppose that the density function* $\bar{f} = \bar{F}'$ *is positive on the open support of* $\bar{F}$. *If*

(15.3) $$\sup_{0<u<1} \frac{u^{\alpha}(1-u)^{\beta}\{\bar{Q}(u) + e^{-1}\}}{\bar{f}(\bar{Q}(u))} < \infty$$

with $0 \le \alpha,\beta < 3/2$, *then*

$$\sup_{0\le u\le 1} |\bar{s}_n(u) - \bar{\Gamma}_n(u)| \overset{a.s.}{=} O(n^{-\tau}),$$

where τ *is as in Theorem* 10.3. *If*

$$0 < \lim_{u\to 0} \frac{u^{\alpha}\{\bar{Q}(u) + e^{-1}\}}{\bar{f}(\bar{Q}(u))} < \infty \quad \text{and} \quad 0 < \lim_{u\to 1} \frac{(1-u)^{\beta}\bar{Q}(u)}{\bar{f}(\bar{Q}(u))} < \infty$$

for some $0 \le \alpha,\beta < 3/2$, *then*

$$\sup_{-e^{-1}\le x\le \bar{\mu}} |\bar{s}_n^{-1}(x) - \bar{\Gamma}_n^{-1}(x)| \overset{a.s.}{=} O(n^{-\tau}),$$

where τ *is as in Theorem* 13.12.

We note that there is another possible definition of the empirical Shannon process and its inverse. The latter would be that we first order the original variables $X_1,\dots,X_n$ to obtain $X_{1:n} \le \dots \le X_{n:n}$ and then apply the transformation $x \to x \log x$. Naturally then the variables $Z_k = X_{k:n} \log X_{k:n}$, $k=1,\dots,n$, will not be ordered, and hence our theory does not hold for the process

$$\frac{1}{n} \sum_{k=1}^{[nu]+1} X_{k:n} \log X_{k:n}, \quad 0 \le u \le 1.$$

Although it would be possible to work out a new convergence theory for such a porcess, we do not attempt this here.

15.3. The empirical redundancy process

Lacking sufficient motivation in doing so, we did not consider the scaled versions of the Lorenz processes of order ν. Of course the derivation of the corresponding results for these processes would be straightforward from Theorem 15.1. On the other hand, there is a motivation to consider the following specially scaled empirical Shannon function:

$$\bar{R}_n(u) = \begin{cases} \dfrac{1}{n\bar{X}_n} \displaystyle\sum_{k=1}^{[nu]+1} Z_{k:n} - \log \bar{X}_n, & 0 \le u \le 1, \\ \dfrac{1}{n} \displaystyle\sum_{k=1}^{n} \dfrac{X_k}{\bar{X}_n} \log \dfrac{X_k}{\bar{X}_n}, & u = 1. \end{cases}$$

The random variable $\bar{R}_n(1)$ was recently considered by Chandra, DeWet and Singpurwalla (1982) who termed it the empirical redundancy. They used it to estimate the "redundancy" of F:

$$\bar{R}_F(1) = E \frac{X}{\mu} \log \frac{X}{\mu}, \quad \mu = EX.$$

Accordingly, we introduce the redundancy function

$$\bar{R}_F(u) = \frac{1}{\mu} \int_0^u \bar{Q}(y)\,dy - \log \mu, \quad 0 \le u \le 1,$$

of F, where $\bar{Q}$ and $Z_{k:n}$ are as in the preceding subsection, and the empirical redundancy process

$$\bar{r}_n(u) = n^{1/2}\{\bar{R}_n(u) - \bar{R}_F(u)\}, \quad 0 \le u \le 1.$$

The approximating copies of the limiting Gaussian process will be

$$\bar{P}_n(u) = \frac{1}{\mu} \bar{\Gamma}_n(u) + \left\{ \frac{\bar{S}_F(u)}{\mu^2} + \frac{1}{\mu} \right\} \int_0^1 B_n(y)\,dQ(y), \quad 0 \le u \le 1,$$

where $\bar{S}_F$ is of (15.1) and $\bar{\Gamma}_n$ is of (15.2).

THEOREM 15.3. (i) _If_ $\bar{\mu} = EX \log X < \infty$, _then_

$$\sup_{0 \le u \le 1} |\bar{R}_n(u) - \bar{R}_F(u)| \xrightarrow{a.s.} 0.$$

(ii) _If_ $\bar{Q}$ _is continuous on_ $(0,1)$, $E(X \log X)^2 < \infty$, _then_

$$\sup_{0 \le u \le 1} |\bar{r}_n(u) - \bar{P}_n(u)| \xrightarrow{P} 0.$$

(iii) _If the density function_ $\bar{f} = \bar{F}'$ _is positive on the open support of_ $\bar{F}$ _and condition_ (15.3) _is satisfied for some_ $0 \le \alpha, \beta < 3/2$, _then_

$$\sup_{0 \le u \le 1} |\bar{r}_n(u) - \bar{P}_n(u)| \overset{a.s.}{=} O(n^{-\tau}),$$

where τ _is as in Theorem_ 10.3.

Proof. Since

$$\bar{R}_n(u) - \bar{R}_F(u) = \frac{1}{\bar{X}_n}(\bar{S}_n(u) - \bar{S}_F(u)) + \bar{S}_F(u)\left(\frac{1}{\bar{X}_n} - \frac{1}{\mu}\right) + \log \mu - \log \bar{X}_n ,$$

(i) follows at once from (i) of Theorem 15.2. Applying the one-term Taylor formula for the difference of the two logarithms, (ii) follows from (ii) of Theorem 15.2, while the two-term formula yields (iii) via (iii) of Theorem 15.2.

16. INDICES OF INEQUALITY, DIVERSITY, AND CONCENTRATION

In connection with total time on test processes the cumulative total time on test transform and its empirical version has received considerable interest in probabilistic and statistical papers in reliability theory. Also, the Gini index and some related measures of economic inequality and industrial concentration have been in use for a long time. For example, Good (1982) has recently pointed out, referring to Keynes (1921, pp. 398-399) that the Gini index was used before Gini (1912) by Lexis (1879) as a "quadratic index of homogenity". The literature on the economic applications of the latter indices is so vast that we could not even attempt to mention all the relevant references. Below we list some commonly used indices, together with some new possibilities, and only list some relevant references we know of in connection with these indices without discussing the contents of these references. We give results on strong consistency and asymptotic normality. Some of these results have been known, or must have been known, in some special cases, or in general as well, perhaps under more stringent regularity conditions. Our versions are obtained as very simple applications of the so far achieved results, and we are not going to compare them to earlier ones.

Cumulative total time on test. (Lexis (1879), Marshall and Proschan (1965), Barlow (1968), Barlow and Doksum (1972), Barlow and Campo (1975), Barlow and Proschan (1975), Barlow (1977), Chandra and Singpurwalla (1978), Langberg, León, Proschan (1980)). We define the cumulative total time on test index of F as

$$I_F^{(1)} = \int_0^1 \frac{N_F^1(y)}{N_F^1(1)}\, dy = \frac{1}{\mu}\int_0^1 N_F^1(y)\, dy \ ,$$

where $N_F^1(\cdot)$ is the total time on test function from the first failure on as in Section 9. Note that if $Q(0) = 0$ then

$$I_F^{(1)} = \int_0^1 \frac{H_F^{-1}(y)}{H_F^{-1}(1)}\, dy \ ,$$

where H_F^{-1} is the total time on test function. The empirical counterpart is

$$I_n^{(1)} = \frac{1}{\bar{X}_n}\int_0^1 N_n^1(y)\, dy$$

$$= \frac{1}{\bar{X}_n} \frac{1}{n^2} \sum_{k=1}^{n} (2(n-k+1)-1) X_{k:n} .$$

Strong consistency, asymptotic normality and convergence rate results would follow from Theorem 9.1 but because of the relation of this index to the cumulative Lorenz index, first noticed by Chandra and Singpurwalla, these results will follow below under milder conditions. The reference to Lexis (1879) will be clear below.

Cumulative Lorenz index (Chandra and Singpurwalla (1978)):

$$I_F^{(2)} = \int_0^1 L_F(y)\,dy$$

$$= \frac{1}{2} I_F^{(1)}$$

with its sample form

$$I_n^{(2)} = \int_0^1 L_n(y)\,dy$$

$$= \frac{1}{\bar{X}_n} \frac{1}{n^2} \sum_{k=1}^{n} (n-k+1) X_{k:n} .$$

Gini index (Gini (1912), Wold (1935), Nair (1936), Kendall (1943), Lomnicki (1952), Taguchi (1968), Gastwirth (1971, 1972), Bhargava and Uppuluri (1975), Goldie (1977), Chandra and Singpurwalla (1978), Sendler (1979), and further references in these papers):

$$I_F^{(3)} = \frac{\frac{1}{2} - \int_0^1 L_F(y)\,dy}{\frac{1}{2}}$$

$$= 1 - 2I_F^{(2)} .$$

The estimator is then

$$I_n^{(3)} = \frac{\frac{1}{2} - \int_0^1 L_n(y)\,dy}{\frac{1}{2}}$$

$$= 1 - 2I_n^{(2)} .$$

Goldie index. This is a measure of cumulative concentration as the area below Goldie's concentration curve:

$$I_F^{(4)} = \int_0^1 L_F^{-1}(y)\,dy$$

$$= 1 - I_F^{(2)} .$$

Its empirical version is

$$I_n^{(4)} = \int_0^1 L_n^{-1}(y)\,dy$$

$$= \frac{1}{\bar{X}_n}\frac{1}{n^2}\sum_{k=1}^{n}(k-1)X_{k:n}$$

$$= 1 - I_n^{(2)}.$$

This was not considered by Goldie (1977), but he clearly deserves these quantities to be named after him. Of course, one may consider, as an imitation of the Gini index in the inverse domain,

$$1 - I_F^{(3)} = \frac{\int_0^1 L_F^{-1}(y)\,dy - \frac{1}{2}}{\frac{1}{2}}$$

as a <u>Lexis index</u> as pointed out by Keynes (1921) and Good (1982), but this is the cumulative total time on test. We have

$$I_n^{(1)} = 2I_n^{(2)} - \frac{1}{n}.$$

<u>Mean difference</u> (Kendall and Stuart (1977), Goldie (1977)):

$$I_F^{(5)} = E|X_1 - X_2| = \int_{-\infty}^{\infty}\int_{-\infty}^{\infty}|x-y|\,dF(x)\,dF(y)$$

$$= 2\mu I_F^{(3)} = 2\mu - 4\mu I_F^{(2)}.$$

The empirical mean difference is

$$I_n^{(5)} = \int_{-\infty}^{\infty}\int_{-\infty}^{\infty}|x-y|\,dF_n(x)\,dF_n(y)$$

$$= \frac{1}{n^2}\sum_{i=1}^{n}\sum_{j=1}^{n}|X_i - X_j|$$

$$= \frac{2}{n^2}\sum_{k=1}^{n}(2k-n-1)X_{k:n}$$

$$= 2\bar{X}_n - 4\bar{X}_n\, I_n^{(2)} + 2n^{-1}\,\bar{X}_n,$$

where F_n is the empirical distribution function of $X_1,\ldots,X_n$.

THEOREM 16.1. (i) <u>If</u> $\mu < \infty$, <u>then</u> $I_n^{(k)} \xrightarrow{a.s.} I_F^{(k)}$, $k=1,\ldots,5$.

(ii) <u>If</u> $Q = F^{-1}$ <u>is continuous on</u> $[0,1)$ <u>and</u> $EX^2 < \infty$, <u>then the distributions of</u> $n^{\frac{1}{2}}(I_n^{(k)} - I_F^{(k)})$ <u>converge to the normal</u> $N(0, v_k^2)$ <u>distribution</u>, $k=1,\ldots,5$, <u>where for the variances we have</u>

$$v_2^2 = v_4^2 = \int_0^1\int_0^1 \sigma_4(s,t)\,ds\,dt$$

$$v_1^2 = v_3^2 = 4v_2^2 ,$$

and

$$v_5^2 = 4(E|X_1-X_2||X_1-X_3| - (E|X_1-X_2|)^2),$$

where $\sigma_4(s,t)$ is the limiting covariance function of the empirical Lorenz process as given in Section 12.

Proof. For $k = 2$, parts (i) and (ii) follow from Theorems 11.1 and 11.2 by noting that the functional $\ell(\cdot) \to \int_0^1 \ell(y)dy$ is continuous with respect to the supremum distance. Of course, $v_2^2 = E(\int_0^1 \Lambda_F(y)dy)^2$. The limiting variance for the mean difference $(k=5)$ is obtained via a lengthy computation of the involved integrals and agrees with the limit of the exact variance of $n^{\frac{1}{2}}(I_n^{(5)} - I_F^{(5)})$ as given in Kendall and Stuart (1977).

Gastwirth index (Gastwirth (1972), Goldie (1977)):

$$\begin{aligned} I_F^{(6)} &= \frac{\int_0^\infty |x-\mu|dF(x)}{2\mu} \\ &= \sup_{0\le u\le 1} |L_F(u)-u| \\ &= F(\mu) - L_F(F(\mu)), \end{aligned}$$

where the last equality holds under our assumption that F is continuous. Here we define its sample version as

$$\begin{aligned} I_n^{(6)} &= \frac{\int_0^\infty |x-\bar{X}_n|dF_n(x)}{2\bar{X}_n} \\ &= F_n(\bar{X}_n) - \frac{1}{\bar{X}_n}\int_0^{\bar{X}_n} x dF_n(x) \\ &= \frac{1}{2\bar{X}_n}\,\frac{1}{n}\sum_{k=1}^n |X_k-\bar{X}_n| . \end{aligned}$$

THEOREM 16.2. (i) If $\mu < \infty$, then $I_n^{(6)} \xrightarrow{a.s.} I_F^{(6)}$.

(ii) If $EX^2 < \infty$ and F is continuously differentiable in a neighbourhood of μ, then

$$\lim_{n\to\infty} pr\{n^{\frac{1}{2}}(I_n^{(6)}-I_F^{(6)}) < x\} = \Phi(x/v_6), \quad -\infty < x < \infty ,$$

where, with $\chi(A)$ denoting the indicator of an event A and $f = F'$,

$$v_6^2 = F(\mu)(1-F(\mu)) + \frac{L_F^2(F(\mu))}{\mu^2} \operatorname{var}\{X\} + \frac{1}{\mu^2} \operatorname{var}\{X\chi(X < \mu)\}$$
$$+ 2 \frac{L_F(F(\mu))}{\mu} \operatorname{covar}\{X, \chi(X < \mu)\}$$
$$- \frac{2}{\mu} \operatorname{covar}\{X\chi(X < \mu), \chi(X < \mu)\}.$$

Proof. (i) Follows from the Glivenko-Cantelli theorem and the strong law of large numbers.

(ii) We have

$$\begin{aligned}
n^{\frac{1}{2}}\{I_n^{(6)} - I_F^{(6)}\} &= n^{\frac{1}{2}}\{F_n(\bar{X}_n) - F(\bar{X}_n)\} \\
&\quad + n^{\frac{1}{2}}\{F(\bar{X}_n) - F(\mu)\} \\
&\quad - \ell_n^*(F(\bar{X}_n)) \\
&\quad + \frac{n^{\frac{1}{2}}}{\mu}\{\int_0^{F(\mu)} Q(y)dy - \int_0^{F(\bar{X}_n)} Q(y)dy\} \\
&= n^{\frac{1}{2}}\{F_n(\bar{X}_n) - F(\bar{X}_n)\} \\
&\quad + f(\tau_n^{(1)})n^{\frac{1}{2}}(\bar{X}_n-\mu) \\
&\quad - \ell_n^*(F(\bar{X}_n)) \\
&\quad - \frac{Q(\tau_n^{(2)})}{\mu} n^{\frac{1}{2}}\{F(\bar{X}_n) - F(\mu)\} \\
&= n^{\frac{1}{2}}\{F_n(\bar{X}_n) - F(\bar{X}_n)\} \\
&\quad - \ell_n^*(F(\bar{X}_n)) \\
&\quad + f(\tau_n^{(1)})n^{\frac{1}{2}}\{\bar{X}_n-\mu\}\{1 - \frac{Q(\tau_n^{(2)})}{\mu}\},
\end{aligned}$$

where ℓ_n^* is as in (13.5), and

$$\min(\bar{X}_n,\mu) \le \tau_n^{(1)} \le \max(\bar{X}_n,\mu)$$

and

$$\min(F(\bar{X}_n),F(\mu)) \le \tau_n^{(2)} \le \max(F(\bar{X}_n),F(\mu)).$$

Hence $Q(\tau_n^{(2)}) \xrightarrow{a.s.} \mu$, implying that the third term converges to zero in probability, for the multiplying factor there has a limiting $(N(0,f^2(\mu)\operatorname{var}\{X\}))$ distribution. Therefore by (2.1) and Lemma 13.3 we obtain that

$$n^{\frac{1}{2}}\{I_n^{(6)} - I_F^{(6)}\} - \tilde{I}_n \xrightarrow{P} 0,$$

where

$$\tilde{I}_n = -B_n(F(\mu)) - \Gamma_n^*(F(\mu)),$$

and an easy computation shows that

$$P\{\tilde{I}_n \leq x\} = \Phi(x/v_6), \quad -\infty < x < \infty,$$

for each n.

<u>Rényi index of order ν</u> (Patil and Taillie (1982)):

$$I_F^{(7)} = G_F^{(\nu)}(1) = \frac{EX^{\nu}-1}{\nu-1}, \quad \nu \neq 1.$$

As Patil and Taillie (1982) note, $\log(1+(\nu-1)I_F^{(7)}) = \log EX^{\nu}$ may be interpreted as the theoretical information gain of order ν per unit, introduced by Rényi (1961). Its empirical version is

$$I_n^{(7)} = G_n^{(\nu)}(1) = \frac{1}{\nu-1}\left(\frac{1}{n}\sum_{i=1}^{n} X_i^{\nu} - 1\right).$$

If $\mu^{(\nu)} = EX^{\nu} < \infty$, then

$$I_n^{(7)} \xrightarrow{a.s.} I_F^{(7)}$$

by the law of large numbers. If $\mu^{(2\nu)} = EX^{2\nu} < \infty$, then the central limit theorem implies that $n^{\frac{1}{2}}(I_n^{(7)} - I_F^{(7)})$ is asymptotically normally distributed with limiting mean 0 and variance

$$v_7^2 = \frac{\mu^{(2\nu)} - (\mu^{(\nu)})^2}{(\nu-1)^2}$$

If, moreover, $EX^{3\nu} < \infty$, then by the Berry-Esséen theorem (Hall, 1982, p.6)

$$\sup_{-\infty<x<\infty} |pr\{n^{\frac{1}{2}}(I_n^{(7)} - I_F^{(7)}) \leq x\} - \Phi(x/v_7)| \leq c_o \frac{EX^{3\nu}}{v_7^3} \frac{1}{n^{\frac{1}{2}}},$$

where $0.4097 \leq \frac{\sqrt{10}+3}{6\sqrt{2\pi}} \leq c_o \leq 0.7975$, the lower bound being a classical result of Esséen.

The Rényi index of order 2 is Pearson's χ^2 divergence, while the case $\nu = 1/2$ is known as the Bhattacharyya (1946) divergence.

<u>Shannon index</u> (Theil (1967), Horowitz and Horowitz (1968), Hexter and Snow (1970), Hart (1971), Patil and Taillie (1982)):

$$I_F^{(8)} = EX \log X$$

with its estimator

$$I_n^{(8)} = \frac{1}{n}\sum_{k=1}^{n} X_k \log X_k .$$

<u>Redundancy</u> (Chandra, DeWet, Singpurwalla (1982)):

$$I_F^{(9)} = \bar{R}_F(1) = E\,\frac{X}{\mu}\log\frac{X}{\mu}\,,$$

$$I_n^{(9)} = \bar{R}_n(1) = \frac{1}{\bar{X}_n}\,\frac{1}{n}\sum_{k=1}^{n} X_k \log X_k - \log \bar{X}_n\,.$$

By the same reasons as in the case of the Rényi indices, if $\mu < \infty$, then

$$I_n^{(k)} \xrightarrow{\text{a.s.}} I_F^{(k)}\,, \quad k = 8,9\,,$$

if $E(X \log X)^2 < \infty$, and then

$$\Delta_n^{(k)}(x) = \mathrm{pr}\{n^{\frac{1}{2}}(I_n^{(k)} - I_F^{(k)}) \le x\} - \Phi(x/v_k) \longrightarrow 0, \quad -\infty < x < \infty,$$

$k = 8,9$, where $v_8^2 = \mathrm{var}\,\{X \log X\}$, and

$$v_9^2 = \frac{1}{\mu^2}\,\mathrm{var}\,\{X \log X\} + \left(\frac{EX \log X}{\mu^2} + \frac{1}{\mu}\right)^2 \mathrm{var}\,\{X\}$$
$$- \frac{2}{\mu}\left(\frac{EX \log X}{\mu^2} + \frac{1}{\mu}\right) \mathrm{covar}\,\{X, X \log X\},$$

and if $E(X \log X)^3 < \infty$, then

$$\sup_{-\infty<x<\infty} |\Delta_n^{(k)}(x)| = O(n^{-\frac{1}{2}})\,, \quad k = 8,9.$$

17. BOOTSTRAPPING EMPIRICAL FUNCTIONS

17.1 Introduction to Bootstrap

In our previous sections we have established weak convergence of mean residual life, total time of test, Lorenz and Goldie processes to appropriate Gaussian processes. Apart from a few special cases (cf. Section 8), these limiting processes are functions of the underlying distributions. Consequently when testing for statistical hypotheses for example, one would have to compute the resulting limiting distribution for each F of interest. The same is true when trying to use our results for constructing confidence bands for the theoretical functionals of the said empirical processes. This type of problems can be solved by adapting the bootstrap method, proposed by Efron (1979, 1982), to the present situation. As we will see, the bootstrap method is simply a Monte Carlo simulation determined by the given observations.

We will assume, without loss of generality, that all the random variables and processes introduced so far and later on are on the same probability space (Ω, A, P) (cf. Lemmas 3.1.1, 3.1.2 and 3.1.3 in M. Csörgő (1983)).

Given our random sample $X_1,\ldots,X_n$, let $X_1^{(n)},\ldots,X_m^{(n)}$ be conditionally independent r.v.'s with common distribution function $F_n(x)$. This means that

$$P\{X_1^{(n)} \le x_1,\ldots,X_m^{(n)} \le x_m | X_1,\ldots,X_n\} = \prod_{i=1}^{m} F_n(x_i) \text{ a.s.}$$

Let $F_{m,n}$ denote the empirical distribution function of $X_1^{(n)},\ldots,X_m^{(n)}$, i.e.,

$$F_{m,n}(x) = \frac{1}{m}\#\{1 \le i \le m : X_i^{(n)} \le x\} ,$$

and let $X_{1:m}^{(n)} \le X_{2:m}^{(n)} \le \cdots \le X_{m:m}^{(n)}$ be their order statistics. Using the probability integral transformation $U_1^{(n)} = F(X_1^{(n)}),\ldots,U_m^{(n)} = F(X_m^{(n)})$ we get

$$E_{m,n}(y) = \frac{1}{m}\#\{1 \le i \le m : U_i^{(n)} \le y\} ,$$

the bootstrapped empirical distribution function of the uniform-(0,1) r.v.'s $U_1 = F(X_1),\ldots,U_n = F(X_n)$. We define

$$\alpha_{m,n}(y) = m^{\frac{1}{2}}(E_n(y)-E_{m,n}(y)), \quad 0 \le y \le 1,$$

the bootstrapped uniform empirical process. Let $U_{m,n}(y)$ be the inverse of $E_{m,n}$, the so-called bootstrapped uniform empirical quantile function. Then the corresponding bootstrapped uniform quantile process is

$$u_{m,n}(y) = m^{\frac{1}{2}}(U_{m,n}(y) - U_n(y)), \quad 0 \le y \le 1.$$

Using now our generated r.v.'s $X_1^{(n)}, \ldots, X_m^{(n)}$, we can bootstrap the processes introduced in the first chapter. Let, putting again $X_{0:m}^{(n)} = 0$,

$$H_{m,n}^{-1}(u) = \begin{cases} \frac{1}{m} \sum_{i=1}^{[mu]+1} (m+1-i)(X_{i:m}^{(n)} - X_{i-1:m}^{(n)}), & 0 \le u < 1, \\ \frac{1}{m} \sum_{i=1}^{m} X_{i:m}^{(n)}, & u = 1, \end{cases}$$

$$\bar{X}_{m,n} = \frac{1}{m} \sum_{i=1}^{m} X_i^{(n)} = \frac{1}{m} \sum_{i=1}^{m} X_{i:m}^{(n)},$$

$$L_{m,n}(u) = \begin{cases} \frac{1}{\bar{X}_{m,n}} \frac{1}{m} \sum_{i=1}^{[mu]+1} X_{i:m}^{(n)}, & 0 \le u < 1, \\ 1, & u = 1, \end{cases}$$

$$D_{m,n}^{-1}(u) = \frac{1}{\bar{X}_{m,n}} H_{m,n}^{-1}(u), \quad 0 \le u \le 1,$$

and

$$M_{m,n}(x) = \frac{1}{1-F_{m,n}(x)} \int_x^{\infty} (1-F_{m,n}(t))dt, \quad 0 \le x < \infty,$$

be the corresponding bootstrapped empirical total time on test, mean, Lorenz, scaled total time on test and mean residual life functions. Also, the bootstrapped empirical Goldie concentration function is $L_{m,n}^{-1}$, the inverse of $L_{m,n}$. Now the appropriate bootstrapped processes are:

$$t_{m,n}(u) = m^{\frac{1}{2}}(H_{m,n}^{-1}(u) - H_n^{-1}(u)), \quad 0 \le u \le 1,$$

$$\ell_{m,n}(u) = m^{\frac{1}{2}}(L_{m,n}(u) - L_n(u)), \quad 0 \le u \le 1,$$

$$s_{m,n}(u) = m^{\frac{1}{2}}(D_{m,n}^{-1}(u) - D_n^{-1}(u)), \quad 0 \le u \le 1,$$

$$c_{m,n}(u) = m^{\frac{1}{2}}(L_{m,n}^{-1}(u) - L_n^{-1}(u)), \quad 0 \le u \le 1,$$

and

$$z_{m,n}(x) = m^{\frac{1}{2}}(M_{m,n}(x) - M_n(x)), \quad 0 \le x < \infty.$$

We formulate our bootstrapped results in one global statement.

THEOREM 17.1. We assume

(17.1) $$0 < \liminf_{n\to\infty} (m/n) \le \limsup_{n\to\infty} (m/n) < \infty ,$$

and $J(r) < \infty$ for some $r > 2$.

(i) Suppose that the density function $f = F'$ is continuous and positive on the open support of F. If

$$J = \sup_{0<u<1} \frac{q(u)(1-u)}{f(Q(u))} < \infty$$

for some O'Reilly weight function q, then there are two sequences of Gaussian processes $\{T_m^*(u), S_m^*(v);\ 0 \le u,v \le 1\}$ such that for each m

$$\{T_m^*(u), S_m^*(v);\ 0 \le u,v \le 1\} \overset{\mathcal{D}}{=} \{T(u), S(v);\ 0 \le u,v \le 1\},$$

and, as $m \wedge n \to \infty$,

$$\sup_{0\le u\le 1} |t_{m,n}(u) - T_m^*(u)| \xrightarrow{P} 0 ,$$

$$\sup_{0\le u\le 1} |s_{m,n}(u) - S_m^*(u)| \xrightarrow{P} 0 .$$

(ii) There is a sequence of Gaussian processes $\{Z_m^*(t);\ 0 \le t < \infty\}$ such that for each m

$$\{Z_m^*(t);\ 0 \le t < \infty\} \overset{\mathcal{D}}{=} \{Z(t);\ 0 \le t < \infty\},$$

and, as $m \wedge n \to \infty$,

$$\sup_{0\le t\le T} |z_{m,n}(t) - Z_m^*(t)| \xrightarrow{P} 0$$

if $T < T_F$, and

$$\sup_{0\le t\le T} |(1-F_{m,n}(t))z_{m,n}(t) - (1-F(t))Z_m^*(t)| \xrightarrow{P} 0$$

for any $T > 0$.

(iii) If $Q = F^{-1}$ is continuous on $[0,1)$, then there is a sequence of Gaussian processes $\{\Lambda_m^*(u);\ 0 \le u \le 1\}$ such that for each m

$$\{\Lambda_m^*(u);\ 0 \le u \le 1\} \overset{\mathcal{D}}{=} \{\Lambda(u),\ 0 \le u \le 1\}$$

and, as $m \wedge n \to \infty$,

$$\sup_{0\le u\le 1} |\ell_{m,n}(u) - \Lambda_m^*(u)| \xrightarrow{P} 0.$$

(iv) If for each $\lambda > 1$

$$\limsup_{u \to 0} \frac{Q(u)q(u)}{Q(u/\lambda)} < \infty$$

holds for some O'Reilly weight function q, then there is a sequence of Gaussian processes $\{\Psi_m^*(u);\ 0 \le u \le 1\}$ such that for each m

$$\{\Psi_m^*(u);\ 0 \le u \le 1\} \stackrel{\mathcal{D}}{=} \{\Psi(u);\ 0 \le u \le 1\}$$

and, as $m \wedge n \to \infty$,

$$\sup_{0 \le u \le 1} |c_{m,n}(u) - \Psi_m^*(u)| \xrightarrow{P} 0.$$

In order to illustrate the practical use of the above bootstrap theorems, we deal with the total time on test transform in detail. Let ϕ be a continuous functional with respect to supremum norm. Let us now generate the bootstrapped total time on test process N times independently: $t_{m,n}^{(i)}$, $1 \le i \le N$. Then, by the Glivenko-Cantelli theorem we get, as $N \to \infty$,

(17.2) $$\frac{1}{N} \#\{1 \le i \le N : \phi(t_{m,n}^{(i)}) \le x \mid X_1, \ldots, X_n\}$$
$$\longrightarrow P\{\phi(t_{m,n}) \le x \mid X_1, \ldots, X_n\} \quad \text{a.s.}$$

and uniformly in x for n and m fixed, for almost all realizations of $X_1, \ldots, X_n$. One can easily verify now that by (17.2) we obtain, as $N \to \infty$,

(17.3) $$T_{N,m,n}(x) = \frac{1}{N} \#\{1 \le i \le N : \phi(t_{m,n}^{(i)}) \le x\} \longrightarrow T_{m,n}(x)$$
$$= P\{\phi(t_{m,n}) \le x\} \quad \text{a.s.}$$

and uniformly in x for n and m fixed. By Theorem 6.2

(17.4) $$T_n(x) = P\{\phi(t_n) \le x\} \longrightarrow T(x) = P\{\phi(T) \le x\}, \quad n \to \infty,$$

and by Theorem 17.1

(17.5) $$T_{m,n}(x) \longrightarrow T(x), \quad m \wedge n \to \infty,$$

for all points of continuity of $T(x)$. Consequently by (17.3), (17.4), (17.5) and (17.1), as $N \wedge m \wedge n \to \infty$,

(17.6) $$|T_{N,m,n}(x) - T_n(x)| \to 0 \quad \text{a.s.}$$

for all points of continuity of $T(x)$. One may, however, take the point of view that we are on the conditional probability space generated by $X_1, \ldots, X_n$. Then by (17.2) we need also a conditional version of (17.5). Hence we now give that conditional version of Theorem 17.1.

By writing

$$\xi_n \xrightarrow{P_c} 0 \ , \quad n \to \infty \ ,$$

we mean that ξ_n converges to zero in probability for almost all realizations of $X_1, X_2, \ldots$.

THEOREM 17.2. <u>We assume</u> (17.1) <u>and</u> $J(r) < \infty$ <u>for some</u> $r > 2$.

(i) <u>Suppose that the density function</u> $f = F'$ <u>is continuous and positive on the open support of</u> F . <u>If</u>

$$J = \sup_{0<u<1} \frac{q(u)(1-u)}{f(Q(u))} < \infty$$

<u>for some O'Reilly weight function</u> q , <u>then as</u> $m \wedge n \to \infty$

$$\sup_{0\le u\le 1} |\{t_{m,n}(u)|X_1,\ldots,X_n\} - T_m^*(u)| \xrightarrow{P_c} 0,$$

$$\sup_{0\le u\le 1} |\{s_{m,n}(u)|X_1,\ldots,X_n\} - S_m^*(u)| \xrightarrow{P_c} 0.$$

(ii) <u>We have, as</u> $m \wedge n \to \infty$,

$$\sup_{0\le t\le T} |\{z_{m,n}(t)|X_1,\ldots,X_n\} - Z_m^*(t)| \xrightarrow{P_c} 0$$

<u>if</u> $T < T_F$, <u>and</u>

$$\sup_{0\le t\le T} |\{(1-F_{m,n}(t))z_{m,n}(t)|X_1,\ldots,X_n\}-(1-F(t))Z_m^*(t)| \xrightarrow{P_c} 0$$

<u>for any</u> $T > 0$.

(iii) <u>If</u> $Q = F^{-1}$ <u>is continuous on</u> $[0,1)$, <u>then, as</u> $m \wedge n \to \infty$,

$$\sup_{0\le u\le 1} |\{\ell_{m,n}(u)|X_1,\ldots,X_n\} - \Lambda_m^*(u)| \xrightarrow{P_c} 0 \ .$$

(iv) <u>If for each</u> $\lambda > 1$

$$\limsup_{u\to 0} \frac{Q(u)q(u)}{Q(u/\lambda)} < \infty$$

<u>holds for some O'Reilly weight function</u> q , <u>then, as</u> $n \wedge m \to \infty$,

$$\sup_{0\le u\le 1} |\{c_{m,n}(u)|X_1,\ldots,X_n\} - \psi_m^*(u)| \xrightarrow{P_c} 0.$$

Consequently, we have for example, in all points of continuity of $\mathcal{T}(x)$ that

(17.7) $$P\{\phi(t_{m,n}) \le x|X_1,\ldots,X_n\} \longrightarrow \mathcal{T}(x) \ , \quad m \wedge n \to \infty \ ,$$

for almost all realizations of $X_1, X_2, \ldots$. Thus by (17.2),(17.4) and (17.7) in every point of continuity of $\underset{\sim}{T}(x)$ we have, as $N \wedge m \wedge n \to \infty$, that

$$(17.8) \qquad \left|\frac{1}{N} \#\{1 \leq i \leq N : \phi(t_{m,n}^{(i)}) \leq x | X_1, \ldots, X_n\} - T_n(x)\right| \longrightarrow 0 \text{ a.s.}$$

for almost all realizations of $X_1, X_2, \ldots$.

The statements (17.6) and (17.8) are completely parallel. In the sequel we will discuss the practical use of (17.6). If one should think that his estimation is a conditional one as formulated in (17.2), then one can give the same discussion for almost all realizations of $X_1, X_2, \ldots$.

Let $\alpha \in (0,1)$ be fixed. When testing statistical hypotheses or constructing confidence bands for H_F^{-1}, we need the value $T_n^{-1}(1-\alpha)$. In this case we have, under the conditions of Theorem 6.2,

$$(17.9) \qquad \lim_{n\to\infty} P\{\phi(t_n) \leq T_n^{-1}(1-\alpha)\} = 1-\alpha ,$$

provided $T(x)$ is continuous in a neighbourhood of $T^{-1}(1-\alpha)$. But we do not know T_n^{-1}. However, (17.6) suggests that $T_{N,m,n}^{-1}(1-\alpha)$ should be a suitable estimator for $T_n^{-1}(1-\alpha)$.

COROLLARY 17.3. *If* $T(x)$ *is continuous in a neighbourhood of* $T^{-1}(1-\alpha)$ *and* (17.1) *holds, then, under the conditions of Theorem* 17.1,

$$|T_{N,m,n}^{-1}(1-\alpha) - T_n^{-1}(1-\alpha)| \longrightarrow 0 \quad \text{a.s.}$$

as $N \wedge m \wedge n \to \infty$.

Proof. By (17.5) and (17.6) we have for all x in some neighbourhood of $T^{-1}(1-\alpha)$ that

$$|T_{N,m,n}(x) - T(x)| \longrightarrow 0 \text{ a.s.}$$

as $N \wedge m \wedge n \to \infty$. Hence by Lemma of Horváth (1984b) we have also

$$(17.10) \qquad |T_{N,m,n}^{-1}(1-\alpha) - T^{-1}(1-\alpha)| \longrightarrow 0 \text{ a.s.}$$

By definition, $T^{-1}(1-\alpha)$ is a point of increase of T and hence by continuity of T we get that $T(T^{-1}(1-\alpha)) = 1-\alpha$. Therefore by (17.4) and the proof of Lemma 1.5.6 in Serfling (1980) we have

$$(17.11) \qquad |T_n^{-1}(1-\alpha) - T^{-1}(1-\alpha)| \longrightarrow 0 .$$

Thus (17.10) and (17.11) result in Corollary 17.3.

We note also that (17.11) gives a proof of (17.9) as well.

By Corollary 17.3 and (17.9) we get immediately the next result.

COROLLARY 17.4. *If* $T(x)$ *is continuous in a neighbourhood of*

$T^{-1}(1-\alpha)$, <u>and</u> (17.1) <u>holds, then, under the conditions of Theorem</u> 17.1,

$$P\{\phi(t_n) \leq T^{-1}_{N,m,n}(1-\alpha)\} \longrightarrow 1-\alpha.$$

<u>as</u> $N \wedge m \wedge n \to \infty$.

In order to use Corollary 17.4, we must know that $T(x)$ is continuous.

PROPOSITION 17.5. <u>Under the conditions of Theorem</u> 6.2 <u>the distribution functions</u>

$$P\{\sup_{0\leq u\leq 1} T(u) \leq x\} , \quad P\{\sup_{0\leq u\leq 1} |T(u)| \leq x\}$$

<u>and</u>

$$P\{\int_0^1 w^2(u)T^2(u)\,du \leq x\}$$

<u>are continuous on</u> $(0,\infty)$, <u>if</u> $\int_0^1 w^2(u)\,du < \infty$.

<u>Proof</u>. The conditions ensure that our limit process $\{T(u);\ 0 \leq u \leq 1\}$ is tight. Hence for any given $\varepsilon,\delta > 0$ there exist $0 = u_1 < u_2 < \ldots < u_K = 1$ such that

$$P\{\max_{1\leq i\leq K-1}\ \sup_{u_i\leq u\leq u_{i+1}} |T(u) - T(u_i)| > \delta\} \leq \varepsilon .$$

Also, $ET^2(u_i) > 0$, $1 < i < K$. Therefore by Theorem 1 of Tsirel'son (1975) we have that

$$P\{\max_{1\leq i\leq K} T(u_i) \leq x\}$$

is continuous in $x \in (0,\infty)$. Hence the first two distribution functions of Proposition 17.5 are continuous on $(0,\infty)$.

It follows from the L_2-decomposability of square-integrable processes (Theorem 2, Section 3, Chapter 5 in Gihman and Skorohod (1969)) that

$$\phi_o(t) = E\exp\{it\int_0^1 w^2(u)T^2(u)\,du\}$$

$$= \prod_{k=1}^{\infty}\left(\frac{\lambda_k}{\lambda_k-2it}\right)^{1/2} ,$$

where $\lambda_1, \lambda_2, \ldots$ are the reciprocals of the eigenvalues of the covariance function of $w(u)T(u)$, $\lambda_k > 0$, $k \geq 1$. It is easy to see that

$$|t|^q|\phi_o(t)|$$

is bounded on the whole line for any $q > 0$. This implies

(Corollary 11.6.1 in Kawata (1972)) continuity of the third distribution function of Proposition 17.5.

REMARK 17.6. The results of this section hold true also for scaled total time on test, Lorenz, mean residual life and Goldie concentration processes with the appropriate changes in the notation.

17.2. Technical Tools

Just like when proving our weak invariance principles for t_n, s_n, z_n, ℓ_n and c_n it is again enough to study only the bootstrapped versions of the corresponding uniform processes. This is on account of immediate statements like

$$P\{\sup_{0\leq y\leq 1} |H_{m,n}^{-1}(y) - \int_0^{U_{m,n}(y)} (1-E_{m,n}(u))dQ(u)| = 0\} = 1$$

for each m and n. The latter is parallel with (6.1) for H_n^{-1}. Our aim is to prove the bootstrapped versions of the weak convergence results of Sections 2, 3, and 4, amounting to Theorems 17.1 and 17.2.

Our method of proof is based on the Bickel and Freedman (1981), and Shorack (1982) representation of bootstrapped uniform empirical and quantile processes (cf. also Bretagnolle (1983)).

Let $\xi_1,\xi_2,\ldots,$ and $\eta_1,\eta_2,\ldots$ be two independent sequences of uniform-(0,1) r.v.'s with corresponding empirical distribution functions

$$\tilde{E}_m(s) = \frac{1}{m} \#\{1 \leq i \leq m : \xi_i \leq s\}$$

and

$$\hat{E}_n(s) = \frac{1}{n} \#\{1 \leq i \leq n : \eta_i \leq s\},$$

based on the first m and n of these two sequences. Let $\xi_{1:m} \leq \cdots \leq \xi_{m:m}$ and $\eta_{1:n} \leq \cdots \leq \eta_{n:n}$ be their respective order statistics, and $\tilde{U}_m(s) = \tilde{E}_m^{-1}(s)$ and $\hat{U}_n(s) = \hat{E}_n^{-1}(s)$ their empirical quantile functions. The corresponding uniform empirical and quantile processes are

$$\tilde{\alpha}_m(y) = m^{\frac{1}{2}}(y - \tilde{E}_m(y))$$

$$\hat{\alpha}_n(y) = n^{\frac{1}{2}}(y - \hat{E}_n(y)),$$

$$\tilde{u}_m(y) = m^{\frac{1}{2}}(\tilde{U}_m(y) - y),$$

and

$$\hat{u}_n(y) = n^{\frac{1}{2}}(\hat{U}_n(y) - y).$$

The already mentioned Bickel and Freedman (1981), and Shorack (1982) representation of bootstrapped uniform empirical and quantile processes representation is as follows.

LEMMA 17.7. <u>For each</u> m <u>and</u> n <u>we have</u>

$$\{E_{m,n}(s),\ U_{m,n}(t),\ \alpha_{m,n}(x),\ u_{m,n}(y);\ 0 \le s,t,x,y \le 1\}$$

$$\stackrel{\mathcal{D}}{=} \{\tilde{E}_m(\hat{E}_n(s)),\ \hat{U}_n(\tilde{U}_m(t)),\ \tilde{\alpha}_m(\hat{E}_n(x)),\ m^{\frac{1}{2}}(\hat{U}_n(\tilde{U}_m(y))-\hat{U}_n(y));$$

$$0 \le s,t,x,y \le 1\}.$$

The proof of this lemma is routine, and follows by lengthy elementary calculations. We omit these details.

The next theorem is a bootstrapped version of the Komlós, Major and Tusnády (1975) and the M. Csörgő and Révész (1978) approximation inequalities.

THEOREM 17.8. <u>Assume</u> (17.1). <u>Then we can define a sequence of Brownian bridges</u> $\{B_m^*(s);\ 0 \le s \le 1\}$ <u>such that</u>

$$P\{\sup_{0\le s\le 1} |\alpha_{m,n}(s)-B_m^*(s)| > A_1 n^{-1/4}(\log n)^{3/4}\} \le B_1 n^{-\varepsilon}$$

<u>and</u>

$$P\{\sup_{0\le s\le 1} |u_{m,n}(s)-B_m^*(s)| > A_2 n^{-1/4}(\log n)^{3/4}\} \le B_2 n^{-\varepsilon}$$

<u>for any</u> $\varepsilon > 0$, <u>where</u> $A_1 = A_1(\varepsilon)$, $A_2 = A_2(\varepsilon)$ <u>and</u> B_1, B_2 <u>are constants</u>.

<u>Proof</u>. Using the Komlós, Major and Tusnády (1975) construction we can define a sequence of Brownian bridges $\tilde{B}_m$, which are independent of $\eta_1,\eta_2,\ldots,$ such that

$$\text{(17.12)}\quad P\{\sup_{0\le s\le 1} |\tilde{\alpha}_m(s)-\tilde{B}_m(s)| > A_{11} n^{-1/2} \log n\} \le B_{12} n^{-\varepsilon}$$

for any $\varepsilon > 0$ and some constants $A_{11} = A_{11}(\varepsilon)$, B_{12}. From the Dvoretzky, Kiefer and Wolfowitz (1956) inequality we get

$$\text{(17.13)}\quad P\{\sup_{0\le s\le 1} |\hat{E}_n(s)-s| > A_{12} n^{-1/2}(\log n)^{1/2}\} \le B_{12} n^{-\varepsilon}$$

for some A_{12} depending on $\varepsilon > 0$. By Theorem 2.C of Burke <u>et al</u>. (1981) (cf. Lemma 1.1.1 of M. Csörgő and Révész (1981)) we have

$$\text{(17.14)}\quad P\{\sup_{0<s\le 1-h_n}\ \sup_{0\le t\le h_n} |\tilde{B}_m(s)-\tilde{B}_m(s+t)| > A_{13} n^{-1/4}(\log n)^{3/4}\} < B_{13} n^{-\varepsilon}$$

with $h_n = A_{12}n^{-1/2}(\log n)^{1/2}$ and $A_{13} = A_{13}(\varepsilon)$. On combining now (17.1), (17.12), (17.13) and (17.14), we obtain

$$(17.15) \quad P\{\sup_{0\le s\le 1} |\tilde{\alpha}_m(\hat{E}_n(s))-\tilde{B}_m(s)| > A_1 n^{-1/4}(\log n)^{3/4}\} \le B_1 n^{-\varepsilon}$$

for any $\varepsilon > 0$ and some constants $A_1 = A_1(\varepsilon)$, B_1.

Aiming at a similar inequality for the representation of the bootstrapped uniform quantile process, we consider

$$(17.16) \quad m^{1/2}(\hat{U}_n(\tilde{U}_m(s)) - \hat{U}_n(s)) = m^{1/2}(\tilde{U}_m(s)-s) + m^{1/2}(\hat{U}_n(\tilde{U}_m(s))-\tilde{U}_m(s)) + m^{1/2}(s-\hat{U}_n(s)).$$

Further we have

$$m^{1/2}(\tilde{U}_m(s)-s) = m^{1/2}(\tilde{U}_m(s)-\tilde{E}_m(\tilde{U}_m(s))) + m^{1/2}(\tilde{E}_m(\tilde{U}_m(s))-s).$$

It is easy to check that

$$\sup_{0\le s\le 1} m^{1/2}|\tilde{E}_m(\tilde{U}_m(s))-s| \le m^{-1/2}$$

and

$$\sup_{0\le s\le 1} |\tilde{U}_m(s)-s| = \sup_{0\le s\le 1} |\tilde{E}_m(s)-s| .$$

Therefore by the Dvoretzky, Kiefer and Wolfowitz (1956) inequality we obtain

$$(17.17) \quad P\{\sup_{0\le s\le 1} |\tilde{U}_m(s)-s| > A_{21}m^{-1/2}\log m\} \le B_{21}m^{-\varepsilon}$$

for some A_{21} depending on $\varepsilon > 0$. Similarly to (17.14) we have again

$$(17.18) \quad P\{\sup_{0\le s\le 1-\tilde{h}_m} \sup_{0\le t\le \tilde{h}_m} |\tilde{B}_m(s)-\tilde{B}_m(s+t)| > A_{22}m^{-1/4}(\log m^{3/4}\} \le B_{22}m^{-\varepsilon}$$

with $\tilde{h}_m = A_{21}m^{-1/2}(\log m)^{1/2}$ and $A_{22} = A_{22}(\varepsilon)$. Thus we proved that

$$(17.19) \quad P\{\sup_{0\le s\le 1} |\tilde{u}_m(s)-\tilde{B}_m(s)| > A_{23}m^{-1/4}(\log m)^{3/4}\} \le B_{23}m^{-\varepsilon}$$

for any $\varepsilon > 0$ and some constants $A_{23} = A_{23}(\varepsilon)$, B_{23}. Next by the M. Csörgő and Révész (1978) approximation we can define a sequence of Brownian bridges $\hat{B}_n$ such that

$$(17.20) \quad P\{\sup_{0\le s\le 1} |\hat{u}_n(s)-\hat{B}_n(s)| > A_{24}n^{-1/2}\log n\} \le B_{24}n^{-\varepsilon}$$

for any $\varepsilon > 0$ and some constants $A_{23} = A_{23}(\varepsilon)$, B_{24}. Combining now (17.20) with (17.17) and with a (17.18) like estimation of the increments of $\hat{B}_n$ we obtain that

$$(17.21)\quad P\{\sup_{0\le s\le 1} |n^{\frac{1}{2}}(\hat{U}_n(\tilde{U}_m(s))-\tilde{U}_m(s))+\hat{B}_n(s)| > A_{25}n^{-1/4}(\log n)^{3/4}\} \le B_{25}n^{-\varepsilon} .$$

Thus, by (17.1), (17.20) and (17.21) we get

$$(17.22)\quad P\{\sup_{0\le s\le 1} |m^{\frac{1}{2}}(\hat{U}_n(\tilde{U}_m(s))-\tilde{U}_m(s))+m^{\frac{1}{2}}(s-\hat{U}_n(s))| > A_{26}m^{-1/4}(\log m)^{3/4}\} \le B_{26}n^{-\varepsilon}.$$

Consequently by (17.19) and (17.22) we have

$$(17.23)\quad P\{\sup_{0\le s\le 1} |m^{\frac{1}{2}}(\hat{U}_n(\tilde{U}_m(s))-\hat{U}_n(s))-\tilde{B}_m(s)| > A_2n^{-1/4}(\log n)^{3/4}\} \le B_2n^{-\varepsilon} .$$

Using now the construction of the proof of Lemma 3.1.2 in M. Csörgő (1983) (cf. Lemma 4.4.4 in M. Csörgő and Révész (1981)), the representation of Lemma 17.7, (17.15) and (17.23), we obtain the two inequalities of Theorem 17.8 as claimed.

Let $P_{\alpha_{m,n}}$, $P_{u_{m,n}}$, and P_B be the probability measures induced by the bootstrapped uniform empirical and quantile processes $\alpha_{m,n}$, $u_{m,n}$, and by a Brownian bridge B. These measures are defined on $\mathcal{D}$, where $(D,\mathcal{D})$ is the Skorohod space of functions defined on $[0,1]$. Let ρ be the Prohorov-Lévy distance of two measures. As a corollary to Theorem 17.8 we have

COROLLARY 17.9. _Assume_ (17.1). _Then_

$$\rho(P_{\alpha_{m,n}}, P_B) = O(n^{-1/4}(\log n)^{3/4})$$

and

$$\rho(P_{u_{m,n}}, P_B) = O(n^{-1/4}(\log n)^{3/4}).$$

Proof. By a result of Dudley (1968) we have

$$\rho(P_{\alpha_{m,n}}, P_B) = \rho(P_{\alpha_{m,n}}, P_{B^*_m}) \le \inf_{\delta>0} (\delta + P\{\sup_{0\le s\le 1} |\alpha_{m,n}(s)-B^*_m(s)| > \delta\}.$$

Taking $\varepsilon = 1$ in Theorem 17.8 and $\delta = A_1(1)n^{-1/4}(\log n)^{3/4}$ we get the first statement of Corollary 17.9, and proof of that of its second one is the same.

We note that Corollary 17.9 is parallel with Proposition 4.1 in Bickel and Freedman (1981).

The next lemma is a bootstrapped version of inequalities of Daniels (1945) and Wellner (1978) (cf. Lemma 2.7).

LEMMA 17.10. *Assume* (17.1). *Then*

$$\lim_{\lambda\to\infty} \limsup_{m\wedge n\to\infty} P\{ \sup_{1/m\le s\le 1} \frac{E_n(s)}{E_{m,n}(s)} > \lambda\} = 0 ,$$

$$\lim_{\lambda\to\infty} \limsup_{m\wedge n\to\infty} P\{ \sup_{1/m\le s\le 1} \frac{s}{E_{m,n}(s)} > \lambda\} = 0 ,$$

$$\lim_{\lambda\to\infty} \limsup_{m\wedge n\to\infty} P\{ \sup_{1/m\le s\le 1} \frac{E_{m,n}(s)}{E_n(s)} > \lambda\} = 0 ,$$

$$\lim_{\lambda\to\infty} \limsup_{m\wedge n\to\infty} P\{ \sup_{0\le s\le 1} \frac{E_{m,n}(s)}{s} > \lambda\} = 0,$$

$$\lim_{\lambda\to\infty} \limsup_{m\wedge n\to\infty} P\{ \sup_{U_{1:m}^{(n)}\le s\le 1} \frac{U_n(s)}{U_{m,n}(s)} > \lambda\} = 0,$$

$$\lim_{\lambda\to\infty} \limsup_{m\wedge n\to\infty} P\{ \sup_{U_{1:n}\le s\le 1} \frac{U_{m,n}(s)}{U_n(s)} > \lambda\} = 0,$$

$$\lim_{\lambda\to\infty} \limsup_{m\wedge n\to\infty} P\{ \sup_{U_{1:m}^{(n)}\le s\le 1} \frac{s}{U_{m,n}(s)} > \lambda\} = 0$$

and

$$\lim_{\lambda\to\infty} \limsup_{m\wedge n\to\infty} P\{ \sup_{1/m\le s\le 1} \frac{U_{m,n}(s)}{s} > \lambda\} = 0 .$$

Proof. These statements follow immediately from Lemma 17.7 and Lemma 2.7.

The next theorem concerns the weak convergence of bootstrapped uniform empirical and quantile processes in weighted sup-norm metrics. These results were first obtained by Lohse (1984), and they parallel Lemma 2.1.

THEOREM 17.11. *If* (17.1) *is satisfied and* q *is an O'Reilly weight function, then with the Brownian bridges* B_m^* *of Theorem* 17.8 *we have, as* $m \wedge n \to \infty$,

$$\sup_{0\le s\le 1} \frac{|\alpha_{m,n}(s)-B_m^*(s)|}{q(s)} + \sup_{1/m\le s\le 1-1/m} \frac{|u_{m,n}(s)-B_m^*(s)|}{q(s)} \xrightarrow{P} 0 .$$

Proof By Theorem 17.8 we get that

(17.24) $$\sup_{\varepsilon \le s \le 1-\varepsilon} |\alpha_{m,n}(s) - B_m^*(s)|/q(s) \xrightarrow{P} 0$$

for any $0 < \varepsilon < 1/2$. Since q is an O'Reilly weight function, we have

(17.25) $$\lim_{\varepsilon \to 0} P\{\sup_{0 \le s \le \varepsilon} |B_m^*(s)|/q(s) > \lambda\} = 0 ,$$

for every $\lambda > 0$ and m. Using now the fact that if q is an O'Reilly function then $q(\lambda s)$ is also an O'Reilly weight function with every $\lambda > 0$, we obtain from Lemmas 17.7, 17.10 and 2.1 that

(17.26) $$\lim_{\varepsilon \to 0} \limsup_{m \to \infty} P\{\sup_{0 \le s \le \varepsilon} |\alpha_{m,n}(s)|/q(s) > \lambda\} = 0$$

for all $\lambda > 0$. Consequently by (17.24), (17.25) and (17.26) we get

$$\sup_{0 \le s \le 1} |\alpha_{m,n}(s) - B_m^*(s)|/q(s) \xrightarrow{P} 0.$$

The proof for the bootstrapped uniform quantile process is the same, and the details are omitted.

The last one of our technical tools is the weak convergence of integrals of the bootstrapped empirical process. In this case we cannot directly follow the proof of Lemma 3.2, for it is based on some martingale properties of the uniform empirical process. We will use Theorem 17.11 instead, and will assume a little bit more than $EX^2 < \infty$. Let

$$\beta_{m,n}(u) = \int_0^u \alpha_{m,n}(v)\,dQ(v).$$

THEOREM 17.12. _If_ (17.1) _is satisfied and_ $J(r) < \infty$ _for some_ $r > 2$, _then_

$$\sup_{0 \le u \le 1} |\beta_{m,n}(u) - \int_0^u B_m^*(v)\,dQ(v)| \xrightarrow{P} 0.$$

Proof. By Theorem 17.11 we have

$$\begin{aligned} \sup_{0 \le u \le 1} &|\beta_{m,n}(u) - \int_0^u B_m^*(v)\,dQ(v)| \\ &\le \sup_{0 \le u \le 1} \int_0^u |\alpha_{m,n}(v) - B_m^*(v)|\,dQ(v) \\ &= o_P(1) \int_0^1 q(v)\,dQ(v) \end{aligned}$$

for every O'Reilly weight function q. Let now $q(t) = (1-t)^{1/r}$. Then

$$J(r) = \int_0^1 (1-t)^{1/r} dQ(t) < \infty,$$

and hence Theorem 17.12 is proved.

The technical tools presented so far are enough for proving Theorem 17.1, combining them appropriately with the proofs of its non-bootstrapped statements. Hence we omit the details. The same can be said about the proof of Theorem 17.2, provided we will have given the conditional versions of the results of this section. Going one by one, we first present the analogue of Lemma 17.7.

LEMMA 17.13. Let m and n be fixed. Then for almost all realizations of $X_1,\ldots,X_n$ and $\eta_1,\ldots,\eta_n$ we have

$$\{E_{m,n}(s), U_{m,n}(t), \alpha_{m,n}(x), u_{m,n}(y);\ 0 \le s,t,x,y \le 1 \mid X_1,\ldots,X_n\}$$

$$\stackrel{\mathcal{D}}{=} \{E_m(\hat{E}_n(s)), \hat{U}_n(\tilde{U}_m(t)), \tilde{\alpha}_m(\hat{E}_n(x)), m^{\frac{1}{2}}(\hat{U}_n(\tilde{U}_m(y)) - \hat{U}_n(y));$$
$$0 \le s,t,x,y \le 1 \mid \eta_1,\ldots,\eta_n\} .$$

From the construction of Theorem 17.8 it is clear that the sequence of Brownian bridges $\{B_m^*;\ m \ge 1\}$ is independent of $\{X_n;\ n \ge 1\}$. Hence the next theorem is true.

THEOREM 17.14. Assume (17.1). Then we have

$$P\{P\{\sup_{0\le s\le 1} |\alpha_{m,n}(s) - B_m^*(s)| > A_1 n^{-1/4}(\log n)^{3/4} \mid X_1,\ldots,X_n\} \le B_1 n^{-\varepsilon}\} \ge 1 - B_1 n^{-\varepsilon}$$

and

$$P\{P\{\sup_{0\le s\le 1} |u_{m,n}(s) - B_m^*(s)| > A_2 n^{-1/4}(\log n)^{3/4} \mid X_1,\ldots,X_n\} \le B_2 n^{-\varepsilon}\} \ge 1 - B_2 n^{-\varepsilon},$$

for any $\varepsilon > 0$, where the Brownian bridges B_m^*, and the constants A_1, A_2, B_1 and B_2 are as in Theorem 17.8.

Theorem 17.14 immediately implies the following results.

THEOREM 17.15. If (17.1) is satisfied and q is an O'Reilly weight function, then with the Brownian bridges B_m^* of Theorem 17.8 we have, as $m \wedge n \to \infty$,

$$\sup_{0\le s\le 1} \frac{|\{\alpha_{m,n}(s) \mid X_1,\ldots,X_n\} - B_m^*(s)|}{q(s)} + \sup_{1/m\le s\le 1-1/m} \frac{|\{u_{m,n}(s) \mid X_1,\ldots,X_n\} - B_m^*(s)|}{q(s)}$$

$$\xrightarrow{P_c} 0 .$$

THEOREM 17.16. *If* (17.1) *is satisfied and* $J(r) < \infty$ *for some* $r > 2$, *then*

$$\sup_{0 \le u \le 1} \left| \{\beta_{m,n}(u) \mid X_1, \ldots, X_n\} - \int_0^u B_m^*(v)\, dQ(v) \right| \xrightarrow{P_c} 0 .$$

These conditional results enable one to prove Theorem 17.2.

18. REFERENCES

ALKER, H.R. Jr. (1965). Mathematics and Politics. MacMillan, New York.

ALLISON, P.D., DE SOLLA PRICE, D., GRIFFITH, B.C., MORAVCSIK, M.J. and STEWART, J.A. (1976). Lotka's law: a problem in its interpretation and application. Social Studies of Science 6, 269-276.

ANDERSON, T.W. and DARLING, D.A. (1952). Asymptotic theory of certain "goodness of fit" criteria based on stochastic processes. Ann. Math. Statist. 23, 193-212.

ATKINSON, A.B. (1970). On the measurement of inequality. J. Economic Theory 2, 244-263.

BARLOW, R.E. (1968). Likelihood ratio tests for restricted families of probability distributions. Ann. Math. Statist. 39, 547-560.

BARLOW, R.E. (1979). Geometry of the total time on test transform. Naval Res. Logist. Quart. 26, 293-402.

BARLOW, R.E., BARTHOLOMEW, D.J., BREMNER, J.M. and BRUNK, H.D. (1972). Statistical Inference under Order Restrictions. Wiley, New York.

BARLOW, R.E. and CAMPO, R. (1975). Total time on test process and applications to failure data analysis. In: Reliability and Fault Tree Analysis (R.E. Barlow, J. Fussell and N.D. Singpurwalla, eds.). 451-482. SIAM, Philadelphia.

BARLOW, R.E. and DOKSUM, K. (1972). Isotonic tests for convex ordering. Proc. Sixth Berkeley Symp. Math. Statist. Probability 1, 293-323.

BARLOW, R.E. and PROSCHAN, F. (1969). A note on a test for monotone failure rate based on incomplete data. Ann. Math. Statist. 40, 595-600.

BARLOW, R.E. and PROSCHAN, F. (1975). Statistical Theory of Reliability and Life Testing: Probability Models. Holt, Rinehart and Winston, New York.

BARLOW, R.E. and PROSCHAN, F. (1977). Asymptotic theory of total time on test processes with applications to life testing. In: Multivariate Analysis (P.R. Krishnaiah, ed.), 227-237. North-Holland, Amsterdam.

BARLOW, R.E. and van ZWET, W.R. (1970). Asymptotic properties of isotonic estimators for the generalized failure rate functions. In: Nonparametric Techniques in Statistical Inference (M.L. Puri, ed.), 159-173. Cambridge University Press.

BERGMAN, B. (1977a). Some graphical methods for maintenance planning. In: Proceedings, 1977 Annual Reliability and Maintainability Symposium, 467-471, Philadelphia.

BERGMAN, B. (1977b). Crossings in the total time on test plot. Scand. J. Statist. 4, 171-177.

BERGMAN, B. (1979). On age replacement and the total time on test concept. Scand. J. Statist. 6, 161-168.

BERGMAN, B. and KLEFSJÖ, B. (1982a). A graphical method applicable to age-replacement problems. IEEE Trans. Reliability 31, 478-481.

BERGMAN, B. and KLEFSJÖ, B. (1982b). TTT-transforms and age replacements with discounted costs. Research Report 1982-4, Dept. of Mathematics, University of Lulea.

BERGMAN, B. and KLEFSJÖ, B. (1984). The total time on test concept and its use in reliability theory. Oper. Res. 32, 596-606.

BHARGAVA, T.N. and UPPULURI, V.R.R. (1975). On an aximatic derivation of Gini diversity, with applications. Metron 33, 1-13.

BHATTACHARYYA, A. (1946). On a measure of divergence between two multinomial populations. Sankhyā 7, 401-406.

BICKEL, P.J. and FREEDMAN, D.A. (1981). Some asymptotic theory for the bootstrap. Ann. Statist. 9, 1196-1217.

BIRNBAUM, Z.W. and MARSHALL, A.W. (1961). Some multivariate Chebyshev inequalities with extensions to continuous parameter processes. Ann. Math. Statist. 32, 687-703.

BRETAGNOLLE, F. (1983). Lois limites du Bootstrap de certaines fonctionnelles. Ann. Inst. Henri Poincaré 19, 281-296.

BRUCKMANN, G. (1969). Einige Bemerkungen zur statistischen Messung der Konzentration. Metrika 14, 183-213.

BRYSON, M.C. and SIDDIQUI, M.M. (1969). Some criteria for ageing. J. Amer. Statist. Assoc. 64, 1472-1485.

BURKE, M.D., CSÖRGŐ, S. and HORVÁTH, L. (1981). Strong approximations of some biometric estimates under random censorship. Z. Wahrscheinlichkeitstheorie verw. Geb. 56, 87-112.

CHANDRA, M., DE WET, T. and SINGPURWALLA, N.D. (1982). On the sample redundancy and a test for exponentiality. Commun. Statist. A - Theory Method. 11, 429-438.

CHANDRA, M. and SINGPURWALLA, N.D. (1978). The Gini index, the Lorenz curve, and the total time on test transform. George Washington University, Serial T-368.

CHANDRA, M. and SINGPURWALLA, N.D. (1981). Relationships between some notions which are common to reliability theory and economics. Math. Oper. Res. 6, 113-121.

CHIANG, C.L. (1960). A stochastic study of the life table and its applications: I. Probability distributions of the biometric functions. Biometrics 16, 618-635.

CHIANG, C.L. (1968). Introduction to Stochastic Processes in Biostatistics. Wiley, New York.

CHIBISOV, D. (1964). Some theorems on the limiting behaviour of empirical distribution functions. Selected Translations Math. Statist. Probability 6, 147-156.

CHUNG, K.L. (1948). On the maximum partial sums of independent random variables. Trans. Amer. Math. Soc. 64, 205-233.

CSÁKI, E. (1977). The law of the iterated logarithm for normalized empirical distribution function. Z. Wahrscheinlichkeitstheorie verw. Geb. 38, 147-167.

CSÖRGŐ, M. (1983). Quantile Processes with Statistical Applications. (Regional Conference Series on Appl. Math.) S.I.A.M. Philadelphia.

CSÖRGŐ, M., CSÖRGŐ, S., HORVÁTH, L. and RÉVÉSZ, P. (1984). On weak and strong approximations of the quantile process. In: Proceedings of the Seventh Conference on Probability Theory (Brasov, Aug. 29 - Sept. 4, 1982), 81-95.

CSÖRGŐ, M. and RÉVÉSZ, P. (1978). Strong approximations of the quantile process. Ann. Statist. 6, 882-894.

CSÖRGŐ, M. and RÉVÉSZ, P. (1980). An estimation of the quantile function via density estimation. Carleton Math. Series No. 169.

CSÖRGŐ, M. and RÉVÉSZ, P. (1981). Strong Approximations in Probability and Statistics. Academic Press, New York-Akadémiai Kiadó, Budapest.

CSÖRGŐ, M. and RÉVÉSZ, P. (1981a). Quantile processes and sums of weighted spacings for composite goodness-of-fit. In: Statistics and Related Topics (M. Csörgő, D.A. Dawson, J.N.K. Rao and A.K.Md. E. Saleh, eds.) 69-87. North-Holland, Amsterdam.

CSÖRGŐ, M. and RÉVÉSZ, P. (1983). Quantile processes for composite goodness-of-fit. In: Colloquia Math. Soc. J. Bolyai 36. Limit Theorems in Probability and Statistics (P. Révész, ed.) 255-304. North-Holland, Amsterdam.

CSÖRGŐ, M., SESHADRI, V. and YALOVSKY, M. (1975). Applications of characterizations in the area of goodness-of-fit. In: Statistical Distributions in Scientific Work, Vol. 2 (G.P. Patil, S. Kotz and J.K. Ord, eds.). 79-90. D. Reidel, Dordrecht.

DALTON, H. (1920). The measurement of the inequality of incomes. Economic Journal 30, 348-361.

DANIELS, H.E. (1945). The statistical theory of the strength of bundles of threads, 1. Proc. Roy. Soc. London Ser. A 183, 405-435.

DASGUPTA, P., SEN, A. and STARRETT, D. (1973). Notes on the measurement of inequality. J. Economic Theory 6, 180-187.

DEHEUVELS, P. (1982). Invariance of Wiener processes and of Brownian bridges by integral transforms and applications. Stochastic Process. Appl. 13, 311-318.

DOKSUM, K.A. and YANDELL, B.S. (1984). Tests for exponentiality. In: Handbook of Statistics Vol. 4 (P. R. Krishnaiah and P.K. Sen, eds.), 579-611. North Holland, Amsterdam.

DUDLEY, R.M. (1968). Distances of probability measures and random variables. Ann. Math. Statist. 39, 1563-1572.

DUGUÉ, D. (1958). Traite de Statistique Theorique et Appliquee. Masson, Paris.

DVORETZKY, A., KIEFER, J. and WOLFOWITZ, J. (1956). Asymptotic minimax character of the sample distribution function and of the multinomial estimator. Ann. Math. Statist. 27, 642-669.

EFRON, B. (1979). Bootstrap methods: another look at the jackknife. Ann. Statist. 7, 1-26.

EFRON, B. (1982). The Jackknife, the Bootstrap and Other Resampling Plans. (Regional Conference Series on Appl. Math.) S.I.A.M. Philadelphia.

EPSTEIN, L. and SOBEL, M. (1953). Life testing. J. Amer. Statist. Assoc. 48, 486-502.

FELLER, W. (1966). An Introduction to Probability Theory and its Applications, Vol. II., Wiley, New York.

GAIL, M.H. and GASTWIRTH, J.L. (1978a). A scale-free goodness-of-fit test for the exponential distribution based on the Lorenz curve. J. Amer. Statist. Assoc. 73, 787-793.

GAIL, M.H. and GASTWIRTH, J.L. (1979b). A scale-free goodness-of-fit test for the exponential distribution based on the Gini index. J. Roy. Statist. Soc. Ser. B. 40, 350-357.

GASTWIRTH, J.L. (1971). A general definition of the Lorenz curve. Econometrica 39 , 1037-1039.

GASTWIRTH, J.L. (1972). The estimation of the Lorenz curve and Gini index. Rev. Economics and Statist. 54, 306-316.

GEFFROY, J. (1958/59). Contributions à la théorie des valeurs extrêmes. Publications de l'Institut de Statistiq des Universités de Paris 7-8, 37-185.

GIHMAN, I.I. and SKOROHOD, A.V. (1969). Introduction to the Theory of Random Processes. W.B. Saunders, Philadelphia.

GINI, C. (1912). Variabilità e mutabilità Contributo allo studio della distribuzioni e relazioni statistiche. Studi Economico-Giuridici della facoltà di Guirisprodenza dell'Università di Cagliari, III parts II.

GOLDIE, C.M. (1977). Convergence theorems for empirical Lorenz curves and their inverses. Adv. Appl. Probability 9, 765-791.

GOOD, I.J. (1982). Comment on Patil and Taillie (1982). J. Amer. Statist. Assoc. 77, 561-563.

GROSS, A.J. and CLARK, V.A. (1975). Survival Distributions: Reliability Applications in the Biomedical Sciences. Wiley, New York.

HALL, M. and TIDEMAN, N. (1967). Measures of concentration. J. Amer. Statist. Assoc. 62, 162-168.

HALL, P. (1982). Rates of Convergence in the Central Limit Theorem. (Research Notes in Mathematics 62). Pitman, Boston.

HALL, W.J. and WELLNER, J.A. (1979). Estimation of mean residual life. Unpublished.

HALL, W.J. and WELLNER, J.A. (1981). Mean residual life. In Statistics and Related Topics (M. Csörgő, D.A. Dawson, J.N.K. Rao and A.K.Md.E. Saleh, eds.). 169-184. North-Holland, Amsterdam.

HART, P.E. (1971). Entropy and other measures of concentration. J. Roy. Statist. Soc. A. 134, 73-85.

HART, P.E. (1975). Moment distributions in economics: an exposition. J. Roy. Statist. Soc. A. 138, 423-434.

HEXTER, J.L. and SNOW, J.W. (1970). An entropy measure of relative aggregate concentrations. Southern Economic Journal 36, 239-243.

HILL, M.O. (1973). Diversity and evenness: a unifying notation and its consequences. Ecology 54, 427-432.

HOEFFDING, W. (1973). On the centering of a simple linear rank statistic. Ann. Statist. 1, 54-66.

HOLLANDER, M. and PROSCHAN, F. (1975). Tests for the mean residual life. Biometrika 62, 585-593.

HOROWITZ, A. and HOROWITZ, I. (1968). Entropy Markov processes and competition in the brewing industry. J. Industrial Economics 16, 196-211.

HOROWITZ, I. (1970). Employment concentration in the Common Market: an entropy approach. J. Roy. Statist. Soc. Ser. A. 133, 463-475.

HORVÁTH, L. (1984a). Strong approximation of renewal processes. Stochastic Process. Appl. 18, 127-138.

HORVÁTH, L. (1984b). Strong approximation of extended renewal processes. Ann. Probability 12, 1149-1166.

HOUSE OF COMMONS (1975). Second Report on Scientific Research in British Universities (First Report from the Select Committee on Science & Technology, Session 1975-76). Cmnd. 504. Her Majesty's Stationery Office, London.

ITÔ, K. and McKEAN, H.P. (1965). Diffusion Processes and their Sample Paths. Springer, New York.

JAESCHKE, D. (1979). The asymptotic distribution of the suprema of the standardized empirical process. Ann. Statist. 7, 108-115.

JAIN, N.C. and PRUITT, W.E. (1975). The other law of the iterated logarithm. Ann. Probability 3, 1046-1049.

JAMES, B.R. (1975). A functional law of the iterated logarithm for weighted empirical distributions. Ann. Probability 3, 762-772.

KAKWANI, N.C. and PODDER, N. (1973). On the estimation of Lorenz curves from grouped observations. Internat. Econom. Rev. 14, 278-292.

KAWATA, T. (1972). Fourier Analysis in Probability Theory. Academic Press, New York.

KENDALL, M.G. (1943). The Advanced Theory of Statistics, Vol. 1, Griffin, London.

KENDALL, M.G. and STUART, A. (1977). The Advanced Theory of Statistics Vol. 1, 4th ed. Griffin, London.

KEYNES, J.M. (1921). A Treatise on Probability. Macmillan, London.

KIEFER, J. (1970). Deviations between the sample quantile process and the sample d.f. In: Nonparametric Techniques in Statistical Inference (M.L. Puri, ed.) 299-319. Cambridge Univ. Press, London.

KLEFSJÖ, B. (1982). On aging properties and total time on test transforms. Scand. J. Statist. 9, 37-41.

KLEFSJÖ, B. (1983a). Some tests against aging based on the total time on test transform. Commun. Statist. - Theor. Math. 12, 907-927.

KLEFSJÖ, B. (1983b). Testing exponentiality against HNBUE. Scand. J. Statist. 10, 65-75.

KOMLÓS, J., MAJOR, P. and TUSNÁDY, G. (1975). An approximation of partial sums of independent r.v.'s and the sample d.f., I. Z. Wahrscheinlichkeitstheorie verw. Geb. 32, 111-131.

KUIPER, N.H. (1960). Tests concerning random points on a circle. Proc. Nederl. Akad. Wetensch. Indag. Math. A. 63, 38-47.

LANGBERG, N.A., LEON, R.V. and PROSCHAN, F. (1980). Characterization of nonparametric classes of life distributions. Ann. Probability 8, 1163-1170.

LEIMKUHLER, F.F. (1967). The Bradford distribution. J. Documentation 23, 197-207.

LEXIS, W. (1879). Über die Theorie der Stabilität statistischer Reiher. Jahrbuch für National Ökonomie und Statistik (1) 32, 604-620. Reprinted in his Abhandlungen zur theorie der Bevolkerungs und Moralstatistik, Jena, 1903.

LOÈVE, M. (1960). Probability Theory. Second ed. Van Nostrand, Princeton.

LOHSE, K. (1984). Zur Konsistenz des Bootstrap-Verfahrens. Dissertation, Hamburg.

LOMNICKI, Z.A. (1952). The standard error of Gini's mean difference. Ann. Math. Statist. 23, 635-637.

MARSHALL, A.W. and PROSCHAN, F. (1965). Maximum likelihood estimation for distributions with monotone failure rate. Ann. Math. Statist. 36, 69-77.

MARSHALL, A.W. and PROSCHAN, F. (1972). Classes of distributions applicable in replacement with renewal theory applications. Proc. Sixth Berkeley Symp. Math. Statist. Probability 1, 395-415. Univ. California Press, Berkeley.

MARTYNOV, G.V. (1978). Omega-Square Criteria (in Russian). Nauka, Moscow.

MASON, D.M. (1982). Some characterizations of almost sure bounds for weighted multidimensional empirical distributions and a Glivenko-Cantelli theorem for sample quantiles. Z. Wahrscheinlichkeitstheorie verw. Geb. 59, 505-513.

NAIR, U.S. (1936). The standard error of Gini's mean difference. Biometrika 28, 428-436.

O'REILLY, N. (1974). On the weak convergence of empirical processes in sup-norm metrics. Ann. Probability 2, 642-651.

PARZEN, E. (1979). Nonparametric statistical data modelling. J. Amer. Statist. Assoc. 58, 13-30.

PATIL, G.P. and TAILLIE, C. (1982). Diversity as a concept and its measurement. J. Amer. Statist. Assoc. 77, 548-567.

PIESCH, W. (1975). Statistische Konzentrationsmaße. Tübinger Wissenschaftliche Abhandlungen, Tubingen.

PYKE, R. and SHORACK, G.R. (1968). Weak convergence of a two-sample empirical process and a new approach to Chernoff-Savage Theorems. Ann. Math. Statist. 39, 755-771.

PROSCHAN, F. and PYKE, R. (1967). Tests for monotone failure rate. Proc. Fifth Berkeley Symp. Math. Statist. Probability, Vol. 3, 293-312. Univ. California Press, Berkeley.

RÉNYI, A. (1961). On measures of entropy and information. Proc. Fourth Berkeley Symp. Math. Statist. Probability. Vol. I. 547-561. Univ. California Press, Berkeley.

ROBBINS, H. and SIEGMUND, D. (1972). On the law of the iterated logarithm for maxima and minima. Proc. Sixth Berkeley Symp. Math. Statist. Probability, Vol. III. 51-70. Univ. California Press, Berkeley.

SEN, A. (1973). On Economic Inequality. Clarendon Press, Oxford.

SEN, A. (1974). Poverty, inequality and unemployment: some conceptual issues in measurement. Sankhyā Ser. C. 36, 67-82.

SENDLER, W. (1979). On statistical inference in concentration measurement. Metrika 26, 109-122.

SENDLER, W. (1982). On functionals of order statistics. Metrika 29, 19-54.

SERFLING, R.J. (1980). Approximation Theorems of Mathematical Statistics. Wiley, New York.

SHORACK. G.R. (1979). Weak convergence of empirical and quantile processes in sup-norm metrics via KMT-construction. Stochastic Process. Appl. 9, 95-98.

SHORACK, G.R. (1982). Bootstrapping robust regression. Comm. Statist. A-Theory Methods 11, 961-972.

SHORACK, G.R. (1982). Kiefer's theorem via the Hungarian construction. Z. Wahrscheinlichkeitstheorie verw. Geb. 61, 369-373.

SHORACK, G.R. and WELLNER, J.A. (1982). Limit theorems and inequalities for the uniform empirical process indexed by intervals. Ann. Probability 10, 639-652.

SKOROHOD, A.V. (1965). Studies in the Theory of Random Processes. Addison-Wesley, Reading, Massachusetts.

STUTE, W. (1982). The oscillation behaviour of empirical processes. Ann. Probability 10, 86-107.

SUGIHARA, G. (1982). Comment on Patil and Taillie (1982). J. Amer. Statist. Assoc. 77, 564-565.

TAGUCHI, T. (1968). Concentration-curve methods and structures of skew populations. Ann. Inst. Statist. Math. 20, 107-141.

THEIL, H. (1967). Economic and Information Theory. North-Holland, Amsterdam.

THOMPSON, W.A., Jr. (1976). Fisherman's luck. Biometrics 32, 265-271.

TSIREL'SON, V.S. (1975). The density of the distribution of the maximum of a Gaussian process. Theory Probab. Appl. 20, 847-856.

VERVAAT, W. (1972). Functional central limit theorems for processes with positive drift and their inverses. Z. Wahrscheinlichkeitstheorie verw. Geb. 23, 245-253.

WEISS, L. (1961). On the estimation of scale parameters. Naval Res. Logist. Quart. 8, 245-256.

WEISS, L. (1963). On the asymptotic distribution of an estimate of a scale parameter. Naval Res. Logist. Quart. 10, 1-9.

WELLNER, J.A. (1978). Limit theorems for the ratio of the empirical distribution function to the true distribution function. Z. Wahrscheinlichkeitstheorie verw. Geb. 45, 73-88.

WELLNER, J.A. (1979). Review of Goldie (1977). Mathematical Reviews 57, #17752.

WILSON, E.B. (1938). The standard deviation of sampling for life expectancy. J. Amer. Statist. Assoc. 33, 705-708.

WOLD, H. (1935). A study on the mean difference, concentration curves and concentration ratio. Metron XII, No. 2, 39-58.

YANG, G.L. (1978). Estimation of a biometric function. Ann. Statist. 6, 112-116.

Lecture Notes in Statistics

Vol. 26: Robust and Nonlinear Time Series Analysis. Proceedings, 1983. Edited by J. Franke, W. Härdle and D. Martin. IX, 286 pages. 1984.

Vol. 27: A. Janssen, H. Milbrodt, H. Strasser, Infinitely Divisible Statistical Experiments. VI, 163 pages. 1985.

Vol. 28: S. Amari, Differential-Geometrical Methods in Statistics. V, 290 pages. 1985.

Vol. 29: Statistics in Ornithology. Edited by B.J.T. Morgan and P.M. North. XXV, 418 pages. 1985.

Vol. 30: J. Grandell, Stochastic Models of Air Pollutant Concentration. V, 110 pages. 1985.

Vol. 31: J. Pfanzagl, Asymptotic Expansions for General Statistical Models. VII, 505 pages. 1985.

Vol. 32: Generalized Linear Models. Proceedings, 1985. Edited by R. Gilchrist, B. Francis and J. Whittaker. VI, 178 pages. 1985.

Vol. 33: M. Csörgő, S. Csörgő, L. Horváth, An Asymptotic Theory for Empirical Reliability and Concentration Processes. V, 171 pages. 1986.